T0318636

Antimicrobial Peptides in Gastrointestinal Diseases

Antimicrobial Peptides in Gastrointestinal Diseases

Edited by

Chi Hin Cho

Department of Pharmacology, School of Pharmacy,
Southwest Medical University, Luzhou, China

ACADEMIC PRESS

An imprint of Elsevier

Academic Press is an imprint of Elsevier
125 London Wall, London EC2Y 5AS, United Kingdom
525 B Street, Suite 1650, San Diego, CA 92101, United States
50 Hampshire Street, 5th Floor, Cambridge, MA 02139, United States
The Boulevard, Langford Lane, Kidlington, Oxford OX5 1GB, United Kingdom

Copyright © 2018 Chi Hin Cho. Published by Elsevier Ltd. All rights reserved.

No part of this publication may be reproduced or transmitted in any form or by any means, electronic
or mechanical, including photocopying, recording, or any information storage and retrieval system,
without permission in writing from the publisher. Details on how to seek permission, further
information about the Publisher's permissions policies and our arrangements with organizations
such as the Copyright Clearance Center and the Copyright Licensing Agency, can be found at our
website: www.elsevier.com/permissions.

This book and the individual contributions contained in it are protected under copyright by the
Publisher (other than as may be noted herein).

Notices
Knowledge and best practice in this field are constantly changing. As new research and experience
broaden our understanding, changes in research methods, professional practices, or medical
treatment may become necessary.

Practitioners and researchers must always rely on their own experience and knowledge in evaluating
and using any information, methods, compounds, or experiments described herein. In using such
information or methods they should be mindful of their own safety and the safety of others, including
parties for whom they have a professional responsibility.

To the fullest extent of the law, neither the Publisher nor the authors, contributors, or editors, assume
any liability for any injury and/or damage to persons or property as a matter of products liability,
negligence or otherwise, or from any use or operation of any methods, products, instructions, or
ideas contained in the material herein.

Library of Congress Cataloging-in-Publication Data
A catalog record for this book is available from the Library of Congress

British Library Cataloguing-in-Publication Data
A catalogue record for this book is available from the British Library

ISBN: 978-0-12-814319-3

For information on all Academic Press publications
visit our website at https://www.elsevier.com/books-and-journals

  Working together
to grow libraries in
developing countries

www.elsevier.com • www.bookaid.org

Publisher: John Fedor
Acquisition Editor: Glyn Jones
Editorial Project Manager: Naomi Robertson
Production Project Manager: Punithavathy Govindaradjane
Cover Designer: Matthew Limbert

Typeset by SPi Global, India

Contents

Contributors

Numbers in parentheses indicate the pages on which the authors' contributions begin.

Robert Bucki (1), Medical University of Bialystok, Bialystok; Jan Kochanowski University in Kielce, Kielce, Poland

Matthew T.V. Chan (21), Department of Anaesthesia and Intensive Care, The Chinese University of Hong Kong, Hong Kong SAR, China

Michelle W. Cheng (35), University of California Los Angeles, Los Angeles, CA, United States

Chi Hin Cho (163), Department of Pharmacology, School of Pharmacy, Southwest Medical University, Luzhou, China

Eduardo R. Cobo (133), University of Calgary, Calgary, Canada

James P. Dilger (77), Department of Anesthesiology, Stony Book Medicine, Stony Brook University, Stony Brook, NY, United States

Bonita Durnaś (1), Jan Kochanowski University in Kielce, Kielce, Poland

Ross J. Fitzgerald (21), Roslin Institute, University of Edinburgh, Easter Bush Campus, Edinburgh, United Kingdom

Lucinda Furci (97), San Raffaele Scientific Institute, Milan, Italy

Tony Gin (21), Department of Anaesthesia and Intensive Care, The Chinese University of Hong Kong, Hong Kong SAR, China

Jeffery Ho (21), Department of Anaesthesia and Intensive Care, The Chinese University of Hong Kong, Hong Kong SAR, China

Ravi Holani (133), University of Calgary, Calgary, Canada

Wei Hu (21), Department of Anaesthesia and Intensive Care, The Chinese University of Hong Kong, Hong Kong SAR, China

John P. Kastelic (133), University of Calgary, Calgary, Canada

Hon Wai Koon (35), University of California Los Angeles, Los Angeles, CA, United States

Ivy K.M. Law (35), University of California Los Angeles, Los Angeles, CA, United States

Li Ma (77), Department of Science Education, Donald and Barbara Zucker School of Medicine at Hofstra/Northwell, Hempstead, NY, United States

Maia S. Marin (133), National Scientific and Technical Research Council (CONICET), Buenos Aires, Argentina

Dermot P.B. McGovern (35), Inflammatory Bowel & Immunobiology Research Institute, Los Angeles, CA, United States

Ewelina Piktel (1), Medical University of Bialystok, Bialystok, Poland

Massimiliano Secchi (97), San Raffaele Scientific Institute, Milan, Italy

Jing Shen (61), Southwest Medical University, Luzhou, China

David Q. Shih (35), Comprehensive Digestive Institute of Nevada, Las Vegas, NV, United States

Tomasz Wollny (1), Holy Cross Oncology Center of Kielce, Kielce, Poland

William K.K. Wu (21), Department of Anaesthesia and Intensive Care, The Chinese University of Hong Kong; Institute of Digestive Diseases, State Key Laboratory of Digestive Diseases, LKS Institute of Health Sciences, CUHK Shenzhen Research Institute, The Chinese University of Hong Kong, Hong Kong SAR, China

Zhangang Xiao (61), Southwest Medical University, Luzhou, China

Lin Zhang (21), Department of Anaesthesia and Intensive Care, The Chinese University of Hong Kong; Institute of Digestive Diseases, State Key Laboratory of Digestive Diseases, LKS Institute of Health Sciences, CUHK Shenzhen Research Institute, The Chinese University of Hong Kong; Laboratory of Molecular Pharmacology, Department of Pharmacology, School of Pharmacy, Southwest Medical University, Luzhou, Sichuan, China

General Introduction of Antimicrobial Peptides and Gastrointestinal Diseases

Chi Hin Cho

Department of Pharmacology, School of Pharmacy, Southwest Medical University, Luzhou, China

This book provides an overview of antimicrobial peptides (AMPs) in particular "Cathelicidin" and "Defensins" in different diseases, such as infections in the upper and lower gastrointestinal (GI) tract. It also covers ulcer repair and inflammation in the stomach and colon and carcinogenesis in the gut. Indeed, AMPs have diversified modes of actions including those on microbes, cancer cells, innate reactions and inflammatory cytokines. All these potential interactions are challenging as these peptides execute opposing reactions, either stimulatory or inhibitory on inflammation and cancer. They are dependent on the stages and microenvironment of diseases and also on the types and levels of the peptides in the targeted tissues and cells. This book provides valuable information on key and pivotal research subjects on the GI tract. No doubt this will be useful to researchers who embark on AMP research by helping them create new research directions and design more effective experimentation.

As this is a short book, the content on the subjects mentioned earlier will be concise and provide a quick and useful reference regarding the various actions for AMPs, not only on the digestive system but also on other organs, such as the lungs and skin. It can also promote research on AMPs and their development as drugs from the bench side to clinical applications for GI diseases.

In this book we have contributors from Asia, Europe, and North America who are working on the front lines of AMP research offering their expert opinions and sharing their rich experiences in the field. Each discusses how AMPs modulate the innate immune system, the mechanisms of antimicrobial action, and how to keep the balance of healthy microbiota and abrogate different pathogens involved in various disorders throughout the GI tract. All these studies open up fundamentally important research directions in studying the host defense mechanisms in relationship with the pathogenesis of different mucosal disorders in the epithelium. Pharmacologically these peptides have diversified and important actions, including antibacteria, antiinflammation, anticancer, and mucosal repair, and further immune-modulatory action in the GI mucosa.

They all mark the important potential for AMPs to be developed as therapeutic agents with multiple targets and mechanisms of action against diseases from the upper to lower GI tract. Therapeutically, all this information will prompt us to find out: (1) better therapeutic agents to induce AMPs production selectively in diseased tissues, (2) more stable and active analogues of AMPs, (3) effective and convenient delivery systems for these peptides in the body, and (4) finally, to make them all pharmaceutically viable for the treatment of various kinds of important and problematic diseases in the GI tract.

Chapter 1

Regulation of Cationic Antimicrobial Peptides Expression in the Digestive Tract

Tomasz Wollny*, Ewelina Piktel†, Bonita Durnaś‡ and Robert Bucki†,‡
*Holy Cross Oncology Center of Kielce, Kielce, Poland, †Medical University of Bialystok, Bialystok, Poland, ‡Jan Kochanowski University in Kielce, Kielce, Poland

Chapter Outline

1 DIGESTIVE TRACT MICROBIOME AS A FACTOR REGULATING EXPRESSION OF CATIONIC ANTIBACTERIAL PEPTIDES

The lumen of the human intestinal tract might be seen as a continuation of the external environment equipped with various antimicrobial mechanisms functioning as part of the host innate and adaptive immune responses [1]. The local defense system named GALT (gut-associated lymphoid tissue) consists of epithelial barriers and unique immune cells and structures located near the epithelium and lamina propria. A single layer of epithelium covered by mucus is composed of enterocytes whose main task is to absorb nutrients from the intestine and by other cell types with highly specialized roles. Goblet cells secrete mucus, Paneth cells are one of several cell types to produce antimicrobial peptides (AMPs) and M cells specialize in sampling and presenting antigens and delivering them to GALT. Together, the epithelium, covered by a thick mucus

Antimicrobial Peptides in Gastrointestinal Diseases. https://doi.org/10.1016/B978-0-12-814319-3.00001-5
Copyright © 2018 Chi Hin Cho. Published by Elsevier Ltd. All rights reserved.

layer and substances such as AMPs and immunoglobulin A (IgA) creates a barrier protecting an internal human body milieu from bacterial infiltration. GALT can be divided into functionally distinct immune inductive and effector sites. The immune inductive sites (secondary lymphoid organs) responsible for the induction phase of the immune response consist of B and T lymphocytes existing as cell follicles or as aggregated forms called Peyer's patches. The immune effector sites include various cells within the lamina propria, such as innate immune cells (macrophages, dendritic cells, and innate lymphoid cells), adaptive effector T cells, IgA-producing plasma cells, as well as the intraepithelial subpopulation of T cells [2,3]. In response to microbial antigens, immune cells are activated to produce various cytokines such as interleukins: IL-4, IL-5, IL-10, IL-13, IL-17, and IL-22, as well as interferons and TNF-α [4].

The intestinal tract is colonized by a large variety of microorganisms— bacteria, fungi, viruses, and parasites that slightly outnumber host cells. Most of the gut bacteria belong to the phyla *Firmicutes* (Gram-positive bacteria, mainly facultative anaerobes, such as *Clostridia, Streptococcaceae, Staphylococcaceae, Enterococcaceae,* or *Lactobacillae*), and *Bacteroidetes* (Gram-negative obligate anaerobes). Representatives of other phyla such as *Proteobacteria, Fusobacteria, Actinobacteria,* and *Verrucomicrobia* are also present in the intestinal tract but in significantly smaller number [3,5,6].

The endogenous gastrointestinal (GI) microbiota could itself be considered as a virtual organ, playing an important role in host health and disease [7]. Gut commensal microorganisms protect against pathogens by competing for nutrients and ecological niches and by producing antimicrobial substances. Commensals are also involved in immune system development, play metabolic functions (absorption of nutrients, vitamins, and short-chain fatty acids production), and maintain the integrity of the intestinal epithelial barrier. In return, the host provides places for commensals to live and allows for nutrient and energy acquisition. So, the relationship between human host and microbiota is mutually beneficial (symbiotic) [8].

The gut microbiota starts to establish early in life (even during gestation), continues after birth, and matures during the course of the first 2 years of life. In a healthy adult, it is quite stable and counts greater than 1000 species and has more than 150 times more genes than the host genome [9].

The majority of bacterial gut strains are present for decades, often for an entire adult life. Most likely, early colonizing microorganisms from parents and other family members are crucial through their metabolic products and impact on the immune system in shaping host health and susceptibility to some diseases [10]. The composition of the intestinal microbiota differs between individuals; however, a high level of diversity is observed at the species level but is low at the phylum level. Differences between stool and mucosa microbiota composition are also observed [5].

The diversity and quantity of gut commensal microorganisms is influenced by endogenous, host-derived factors such as genetic background, age, acid secretion, intestinal motility, immune response, as well as exogenous factors, such as diet, age, stress, antibiotic usage, proton pump inhibitor (PPI) usage, probiotics/prebiotics, and hospitalization [11,12].

Commensal bacteria and the gut immune system influence each other. The microbiota plays an important role in maturation of the immune system, stimulates the release of mucus from goblet cells, activates Paneth cells to secrete antimicrobial peptides, and teaches T-helper cells to be tolerogenic to non-pathogenic antigens and bacteria. A symbiotic relationship between host and microbiome leads to intestinal immune homeostasis. A lack of this balance increases the risk of developing infectious diseases in newborns as well as developing inflammatory, allergic, or some autoimmune diseases (e.g., type 1 diabetes) or obesity in later stages of life [13].

Alterations of the intestinal microbiota are associated with several immune-mediated inflammatory diseases including inflammatory bowel disease, multiple sclerosis, rheumatoid arthritis, systemic lupus erythematosus, psoriasis, and psoriatic arthritis [12,14].

The coevolution of microbiota and host cells established several mechanisms to protect human tissues from microorganisms colonizing body surfaces. A thick inner mucus layer, antimicrobial peptides, and immunoglobulin A minimize the contact of microbiota with the epithelium and act as a first line of defense. If any commensal bacteria invade the epithelium and reach the lamina propria, the second line of defense is activated. Apart from intestinal macrophages, IL-22-producing innate lymphoid cells are crucial elements in preventing the systemic dissemination of commensal gut bacteria [6,15]. Interestingly, commensal microbiota, particularly *Clostridial,* and *Bacteroidetes,* might be involved in the development of resistance to fungal colonization, as recognized by Fan et al., using a mouse model of *Candida albicans* gastrointestinal colonization in antibiotic-treated adult mice [16]. It was demonstrated that the mechanism of such phenomenon includes the activation of HIF-1α, a transcription factor involved in activating innate immune effectors, which is followed by induction of CRAMP (mouse cathelicidin-related AMP). The above results suggest that activation of immune effectors present at mucosal surfaces of the digestive tract might provide a novel therapeutic approach for preventing invasive fungal disease in humans [16].

Commensal bacteria inhabit mainly the outer, less-dense mucus layer where they grow, forming a biofilm structure. The inner epithelium associated layer, generally free of microorganisms, acts as physical barrier between host cells and microbiota [17]. The mucus layer not only separates commensal bacteria from the underlying submucosa but also plays a crucial role in the recognition and the initiation of an innate response against invading pathogens. A large portion of such a defense is assured by continuous production of AMPs. It should be assumed that a specific kind and number of bacteria that colonize the digestive tract require a corresponding panel of AMPs that are definitely involved in maintaining healthy microbiota status. However, the mechanisms of distinguishing commensal from pathogenic bacteria by the intestinal immune system is complicated and not fully understood. It is speculated that it involves lower susceptibility of microbiota species to the killing action of AMPs and/or lower ability of those bacteria to activate AMP transcription. Generally, in the first stage of the immune response associated with AMP production, microbial-associated molecular patterns (MAMPs) and damage-associated molecular patterns (DAMPs)—evolutionarily conserved structures present on microbes—are recognized by toll-like receptors (TLRs) and the nucleotide oligomerization domain-like receptors (NLRs) expressed on epithelial cells and various cells localized within the lamina propria (e.g., dendritic cells, macrophages, granulocytes, lymphocytes, and innate lymphoid cells). In the early innate immune response, mainly antigen-presenting cells (dendritic cells) and phagocytes (macrophages, granulocytes) are involved [18] but the specific molecular patterns involved and mechanisms of cell signaling controlling AMP expression have not yet been elucidated.

The gut epithelium produces various antimicrobial peptides (defensins, cathelicidin LL-37, C-type lectins (REG family), ribonucleases, and others such as calprotectin). The plethora of AMPs is probably associated with the continuous threat of microbial invasion at the intestinal tract. In general, they have a broad spectrum of activity (they rapidly kill or inactivate microorganisms such as bacteria, fungi, viruses, protozoa) and a similar mode of action. Most AMPs kill bacteria through nonenzymatic disruption of the cell-membrane [19]. Because the cell membrane is essential to maintaining cell integrity, the likelihood that a microorganism will develop resistance to such antimicrobials is very low. So, the mechanism of action and the large diversity of these substances produced by the mammalian intestine are probable key to retaining the activity of AMPs over evolutionary timescales [20].

AMPs act as endogenous antibiotics and also can modulate immune responses. They accompany the host from early stages of life. Defensins and lactoferrin are present in the amniotic fluid, where together with endotoxin-neutralizing proteins, lipopolysaccharide-binding protein (LBP), and epidermal growth factors (EGF), protect against pathogenic microbes and excessive immune responses [21]. Throughout the entire lifespan, AMPs control the commensal microbiota by preventing excessive outgrowth of the bacterial population and its translocation to tissues that are considered sterile, they limit bacteria-epithelial cell contact and protect against exogenous pathogens [19].

The role of AMPs in the regulation of the composition of intestinal microbial communities has been demonstrated in a study with two genetic mouse models: mice expressing a human α-defensin and defensin-deficient mice. Although no differences in the total number of colonizing bacteria were observed, there were defensin-dependent changes in the proportions of two main phyla (*Firmicutes* and *Bacteroidetes*) compared with wild-type mice. A decrease in the expression of these molecules can lead to changes in the composition of the commensal gut bacteria resulting in chronic intestinal inflammation [22]. In the study where transgenic mice displayed overexpressing human defensin HD-5, mice exposed to oral challenge with virulent *Salmonella typhimurium* strains showed a protective role of these molecules against intestinal pathogens [23].

2   FACTORS REGULATING EXPRESSION OF AMPs AND MOLECULAR MECHANISMS GOVERNING THIS PROCESS

Due to the continuous contact of surfaces of the digestive tract with the external environment and a constant interaction with a variety of bacteria, viruses, fungi, and parasites, production of a broad spectrum of antimicrobial agents by epithelial surfaces of the intestine is crucial to defend the digestive tract against pathogenic microbes. Accordingly, the mammalian gut epithelium produces a diverse collection of AMPs in both a constitutive and inducible manner. The constitutive epithelial synthesis of some AMPs seems to be a key factor in limiting bacterial adherence and translocation [1]. The expression of intestinal AMPs is high at birth and in early infancy during commensal microbiota colonization. It underlines the crucial role of antimicrobial peptides in shaping immune homeostasis [19]. Interestingly, mounting evidence suggests that the expression and local distribution of AMPs is varied throughout the digestive tract, indicating that their expression is tissue-specific and probably influences the commensal microbiota profile in the different segments of the intestinal tract [24]. Additionally, some AMPs such as LL-37 might increase the resistance of the mucosa to bacterial invasion by increasing cell regeneration and cell mechanical properties (cytoskeletal actin remodeling) that prevent bacteria entrance into epithelial cells. To date, Paneth cells are known to produce lysozyme, sPLA2 (secretory phospholipase A2), and angiogenin 4, and are recognized as the site of exclusive production of α-human defensin 5 and 6 (HD5 and HD6), where defensins are stored as propeptides and further cleaved by trypsin [20,25–27]. β-Defensins are expressed throughout the epithelium in both the small and large intestines [28], but with high variability, particularly for hBD-1, which is expressed at higher levels in the colon, stomach, and ileum than in the duodenum, jejunum, and tongue [24]. Cathelicidin LL-37 expression is highly associated with intestinal microbiota and is highest during infancy until weaning from breastfeeding. In adulthood, cathelicidin expression is high in the colon and ceases in the small intestine [29], but its additional synthesis in epithelial cells might be induced by infectious factors and proinflammatory stimuli [30]. Importantly, AMPs produced by epithelial cells are retained by the surface-overlaying mucus secreted by goblet cells and, in effect, are highly concentrated at or near the epithelial surface. They contribute to the creation of an additional barrier responsible for limiting the amount of bacteria in the vicinity of the epithelium and controlling the translocation of both pathogenic and commensal bacteria through the gut barrier without influencing the abundance of microorganisms found in the lumen [31,32].

It is established that enhanced expression of AMPs might be affected by a number of microbial-dependent and microbial-independent factors, resulting in increase of local and systemic concentration of antimicrobial agents. The control of expression, secretion, and activity of epithelial AMPs is crucial in order to (i) minimize the toxic effect of AMPs against mammalian cells determined by AMPs having a membrane permeabilizing-based mechanism of antimicrobial

activity, (ii) avoid activating unnecessary inflammatory responses due to immunomodulatory properties of some AMPs, and (iii) limit alterations in the intestinal microbial composition, which would have a negative effect on beneficial
contributions of the microbiota. So far, a variety of factors affecting the expression of AMPs in the digestive tract have been identified and thoroughly investigated. This group includes proinflammatory factors such as bacteria-derived
endotoxins LPS and LTA, cytokines, that is, IFN-γ, IL-1, or TNF-α, nutritional
compounds (vitamin D, primary and secondary bile acids, retinoic acid), and
by-products of microbial metabolism (butyrate) [33] (Table 1).

TABLE 1 Factors Regulating AMP Expression in the Gastrointestinal Tract

Antimicrobial Peptide	Regulation Factors/GI Disorders	Effect	Refs.
HD-5 and HD-6	NOD2 activation (MDP, LPS), *Salmonella* infections, ulcerative colitis	↑	[23, 34–36]
	Crohn's disease	↓	[37]
hBD-1	IFN-γ and LPS in monocytes, viral infection	↑	[38]
	Shigella infection, ulcerative colitis, Crohn's disease	↓	[39–41]
hBD-2, 3, 4	bacterial infections (including *Salmonella* spp., *H. pylori*), flagellin of probiotic bacteria, IL-1β, IL-1α, and TNF-α, vitamin D, sulforaphane, sodium butyrate	↑	[42–46]
	Shigella infections	↓	[39]
LL-37/hCAP18	Vitamin D, butyrate, LPS, IL-18, infections (including *H. pylori*), bile acids	↑	[47–52]
	Shigella infections	↓	[39]
Secretory leukocyte proteinase inhibitor (SLPI)	TNF-α, IL-1β	↑	[53]
	HIV, HPV infection, *H. pylori*	↓	[54–56]
Elafin	IL-1, TNF-α	↑	[57]
sPLA2	LPS	↑	[58]
Lysozyme	Ulcerative colitis	↑	[59]
	Salmonella infections	↓	[60]
BPI	LPS, LTA, cell damage	↑	[61,62]

Recent studies focus mainly on the identification of novel pattern-recognition receptors (PRRs) and molecular signaling pathways involved in the initiation of innate immune response against digestive tract-related pathogens. To date, it has been demonstrated that production of cytokines, chemokines, and AMPs in response to microbial stimuli is conditioned by a number of different signaling molecules and cascades, including (MyD)88 (myeloid differentiation primary response gene), MAPK (mitogen-activated protein kinases), and NF-κB (nuclear factor kappa-light-chain-enhancer of activated B cells)-associated pathways [63,64]. TLRs are currently considered the most studied PRRs, which induce intracellular signaling mainly by (i) MyD88 or (ii) toll/interleukin 1 receptor domain-containing adapter inducing IFN-β (TRIF), resulting in activation of the NF-κB-dependent pathway in response to microbial-derived stimuli. Lipopolysaccharide (LPS) and peptidoglycan (PGN) being components of Gram-negative and Gram-positive bacteria, respectively, are known to induce β-defensins through PAMP receptors in the colonic epithelium. Interestingly, both of these molecules do not promote AMP production in the same signaling pathways. The effect of LPS is conditioned by its interaction via TLR4 and its accessory molecule MD-2 in a mechanism dependent on NF-κB/AP-1 and Jun kinase. Apro-stimulatory effect of PGN was promoted via interaction with TLR2 and TLR6 in an NF-κB-dependent manner [65]. Accordingly, low expression of these receptors determines poor response of intestinal cells to LPS and PGN [65]. Although the TRIF pathway has been largely connected with an antiviral response so far, in contrast to MyD88-utilizing TLRs [66], there is a data demonstrating that the TRIF pathway is crucial to activate immune responses against the intestinal pathogen *Yersinia enterocolitica*. Sotolongo et al. confirmed that TRIF-deficient mice were characterized by increased bacterial dissemination after oral infection with *Y. enterocolitica* and had impaired phagocytosis of gram-negative pathogens such as *Y. enterocolitica, S. typhimurium,* and *Escherichia coli* [67].

A number of studies demonstrated an increase of antimicrobial expression and secretion due to interaction of microbial supernatants with cellular receptors of intestinal epithelial cells. Pathogenic *Salmonella enteritidis, S. typhimurium, Salmonella typhi,* and *Salmonella dublin* are known to induce human β-defensin-2 (hBD-2) mRNA expression in Caco-2 human carcinoma cells [64] and MKN45 human gastric cancer cells [68]. Importantly, FliC (flagella filament structural protein), the major flagellar filament protein of *S. enteritidis,* acting via activation of an NF-κB-mediated pathway, was recognized as the main hBD2-inducing factor in bacterial culture supernatants [64]. In a similar manner, flagellin from probiotic bacteria, mainly lactobacilli, upregulate hBD-2 expression by NF-κB/AP-1 and MAPK-dependent pathways [69]. Mixed probiotic components derived from *Lactobacillus* spp. also upregulated the expression of CRAMP in liver macrophages of rats [70].

In contrast, a bacterial supernatant of *Helicobacter pylori* (*H. pylori*) was insufficient to induce hBD-2 from MKN45 cells, indicating that this effect is strain-dependent, and for some bacterial strains, a direct interaction of bacterial cells is crucial to affect the expression of AMPs. Research performed by Wada et al. revealed the implication of genetic factors in regulation of β-defensins, because only *H. pylori* strains characterized by the presence of a *Cag* pathogenicity island (*cag* PAI) were able to increase the hBD-2 mRNA level [68].

Apart from the TLR-mediated stimulation of AMPs, nucleotide-binding domain leucine-rich repeat-containing receptors (NLRs) are another important class of innate immune receptors. This group includes nucleotide-binding oligomerization domain protein-1 (NOD1) and NOD2, which are expressed in Paneth cells and intestinal epithelial cells, respectively. Interestingly, human NOD1 specifically detects a specific ligand characterized exclusively for Gram-negative bacterial PGN, resulting in activation of the transcription factor NF-κB-dependent pathway [71], and constitutes a backup mechanism for activation of intestinal immunity during infection with pathogenic Gram-negative enteropathogens that bypass TLR signaling [72]. In contrast, NOD2 recognizes bacterial muramyl dipeptide common to both Gram-positive and Gram-negative bacteria. Moreover, it was demonstrated that NOD2 is crucial in proper colonization of the intestinal microbiota as demonstrated using NOD2-deficient mouse-based research [73].

An unfavorable phenomenon is suppression of AMPs by some infectious factors. Islam et al. demonstrated that in early stages of *Shigella* spp. infection, production of LL-37 peptide and hBD-2 is significantly downregulated, which provides an escape mechanism for bacterial pathogens [39]. The same targeted survival strategy is utilized by pathogenic *Salmonella* strains in Paneth cells, in which secretion of lysozyme and cryptdin peptides is decreased due to activation of the p38 MAPK pathway [60] and during *Shigella flexneri* infection as a result of blocking gene expression of hBD-1, hBD-3, and CCL20. This compromises the recruitment of dendritic cells to the lamina propria of infected tissues [40].

In similar signaling pathways, AMPs are induced by proinflammatory cytokines, whose production is induced by activated PRRs, providing an additional mechanism for the local control of infection in the gastrointestinal tract. Through the NF-κB-mediated mechanism, such cytokines as IL-1β, IL-1α, and TNF-α induce hBD-2 mRNA expression in Caco-2 and HT-29 human colonic epithelial cells [42]. However, the stimulation of this AMP might also be induced using an NF-κB-independent signaling pathway, because IL-1β and TNF-α combined with dexamethasone are able to induce β-defensin production despite the employment of NF-κB antagonists [74]. IL-18 is also recognized as a factor stimulating the production of LL-37 and α-defensins during *Cryptosporidium parvum* infection of intestinal epithelial cells [47]. Accordingly, expression of cathelicidin LL-37 is increased in inflamed and noninflamed mucosa in patients suffering from ulcerative colitis [48].

Some reports indicate that diet and supplementation with nutritional compounds might be beneficial for improvement and/or modulation of immune function of the GI tract. The most effective supplements on the induction of expression of AMPs are shown to be vitamin D, primary and secondary bile acids, butyrate, and sulforaphane.

Vitamin D, particularly its hormonal form, 1,25-dihydroxyvitamin D(3) $(1,25(OH)_2D_3)$, is considered an immune system modulator affecting the secretion of LL-37 and defensins via direct induction of CAMP and DEFB4 gene expression through VDR (vitamin D receptor)-mediated TLR activation (Fig. 1). The mechanism of this phenomenon proposed by Wang et al. assumes that (i) TLR-signaling activates NF-κB binding which is followed by an increase of VDR, (ii) formation and translocation of ligand-bound VDR: RXR heterodimers into the nucleus occurs in the presence of high levels of $1,25(OH)_2D_3$, and (iii) binding of heterodimer to the VDRE (vitamin D response element) in the promoter of the human CAMP gene [75]. Interestingly, both cathelicidin and vitamin D3 mutually modulate each other's biological activity, because vitamin D enhances phagosome maturation and increases the antimicrobial activity of LL-37 peptide and, simultaneously, cathelicidin serves as a mediator of vitamin D3-induced autophagy, which improves the protection of the host against infections caused by such pathogens as *Mycobacterium tuberculosis* [49].

Research performed by Wang et al. indicated that vitamin D, through a VDRE in the promoter, enhances the production of hBD-2 via induction of DEFB4, a gene for this AMP, but this effect is notably weaker than that observed for cathelicidin LL-37. To date, it has been demonstrated that activation of the vitamin D pathway alone is not sufficient to induce strong expression of hBD-2, thus additional signaling pathways are required [76]. in vitro studies presented that induction of the DEFB4 gene requires, apart from high concentration of vitamin D, TLR activation followed by upregulation of IL-1β and IL-1 receptor and downregulation of IL-1 receptor antagonist [76]. Moreover, 1,25 $(OH)_2D_3$ was shown to strongly induce the expression of NOD2 in epithelial cells and synergistically induced NF-κB-mediated pathways and expression of the hBD-2 gene [77]. Research performed by Schwab et al. also presented that the vitamin D pathway is involved in the induction of hBD-2 mRNA by dietary histone deacetylase (HDAC) inhibitor and sulforaphane (SFN) in Caco-2 and HT-29 colonocytes. In this process, hBD-2 expression is regulated through VDR, MAPK/ERK 1/2 and NF-κB signaling [43].

Bile acids, due to binding to and activation of a variety of nuclear receptors, are considered as other factors affecting the expression of AMPs, which contribute to the sterility of the bile tract. Particularly, chenodeoxycholic acid (CDCA) and ursodeoxycholic acid (UDCA) upregulate cathelicidin expression by (i) binding to the FXR:RXR heterodimer that binds to the VDRE in the CAMP gene promoter and (ii) the activation of the ERK1/2 signaling pathway that, in turn, induces VDR protein expression [50]. In addition, lithocholic acid (LCA), being a by-product of CDCA metabolism by bacteria in the colon, increases human CAMP transcription due to its affinity to FXR, pregnane X receptor

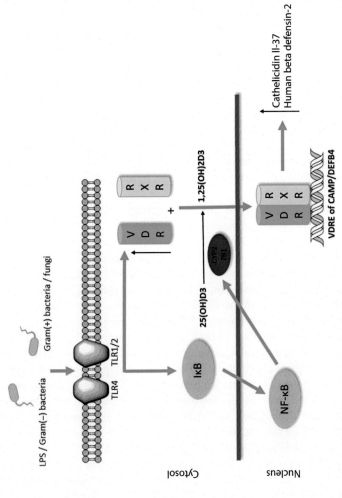

FIG. 1 Vitamin D-mediated induction of CAMP/DEFB4 gene expression.

(PXR), and VDR [78]. LCA also acts synergistically with butyrate (i.e., a short-chain fatty acid produced in the colon by fermentation of dietary fiber by anaerobic bacteria) to induce human CAMP gene expression in human colonic epithelial HT-29 cells [79]. This is particularly important in light of reports suggesting the clinical usefulness of butyrate derivatives in treatment of bacterial infections of the GI tract by inducing LL-37 peptide production [51]. There are also reports demonstrating an increase of hBD-2 mRNA expression in colonocytes and gingival epithelial cells after pretreatment with sodium butyrate [43,44].

3   REGULATION OF AMPs' EXPRESSION IN AUTOIMMUNE DISORDERS ASSOCIATED WITH THE GASTROINTESTINAL TRACT

It is established that AMPs are expressed and secreted throughout the GI tract in order to protect the host from external environment pathogens. Nevertheless, their expression is frequently impaired in GI diseases, which not only blocks the protective role of these proteins but may also be considered as a pathogenic factor for development of GI tract-associated disorders (Fig. 2).

In the oral cavity, a diversity of AMPs protect the cells from oral bacteria such as *Porphyromonas gingivalis, Aggregatibacter actinomycetemcomitans, Streptococcus gordonii, Prevotella intermedia, Fusobacterium nucleatum,* and *Streptococcus sanguinis*, which are associated with the development of periodontal diseases [80]. In addition to that constitutive expression, an increased concentration of hBD-2 was observed in gingivitis and periodontitis in response to developing infection [81,82]. Cytokines, including TNF-α, IL-1β, and IL-17, amplify the secretion of β-defensins, improving the protection against *Candida* colonization in the oral cavity [83]. Human cathelicidin LL-37 in the oral cavity is expressed mainly in inflamed gingival tissues, buccal mucosa, and tongue epithelium, and its increased level is correlated with the progress of gingival sulcus [84]. In contrast, insufficient levels of LL-37 are observed in Papillon-Lefèvre syndrome and Haim-Munk syndrome, which additionally underline the importance of human cathelicidin in the prevention of periodontal disease [85].

Infections of the esophagus are rather uncommon in normal conditions. However, an immunocompromised host very often suffers from viral and fungal infections (like *C. albicans,* CNV or HSV), while bacterial infections are rare [33]. Interestingly, expression of AMPs is high in esophageal tissue; however, it showed weak potency to kill microbiota; this may explain a small resistance to infection of *C. albicans* [86]. In the next study, the authors showed that polymorphonuclear leukocytes (PMNs) reinforced defensin expression in the esophageal epithelium. Therefore, as speculated, patients with neutropenia are not stimulated by PMNs to express antimicrobial peptides, and in consequence, there is high incidence of esophagitis and Candida-related deaths in neutropenic

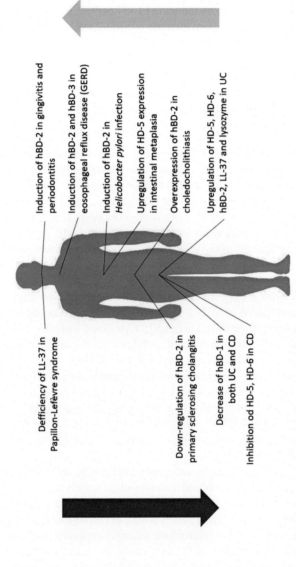

Deffiiciency of LL-37 in
Papillon-Lefèvre syndrome

Induction of hBD-2 in gingivitis and
periodontitis

Induction of hBD-2 and hBD-3 in
eosophageal reflux disease (GERD)

Induction of hBD-2 in
Helicobacter pylori infection

Upregulation of HD-5 expression
in intestinal metaplasia

Overexpression of hBD-2 in
choledocholithiasis

Down-regulation of hBD-2 in
primary sclerosing cholangitis

Upregulation of HD-5, HD-6,
hBD-2, LL-37 and lysozyme in UC

Decrease of hBD-1 in
both UC and CD

Inhibition od HD-5, HD-6 in CD

FIG. 2 Alterations in AMP concentration in the course of gastrointestinal tract diseases.

subjects [87]. In very common gastroesophageal reflux disease (GERD), an induction of beta-defensin expression (hBD-2 and hBD-3) was found, but to a minor degree [45]. However, a relationship between GERD and chronic periodontitis has been found [88]. The authors postulated that hyposalivation in GERD may have an influence on the development of chronic periodontitis by allowing for the proliferation of intraoral bacteria due to diminished production of oral AMPs against *P. gingivalis* and *A. actinomycetemcomitans*.

It is generally accepted that colonization of the stomach by *H. pylori* can result in such GI illnesses as chronic gastritis, peptic ulcers, and gastric cancer [89,90]. The discovery of this Gram-negative bacterium and its related morbidity has provoked much interest in the role of AMPs in the stomach. It has been shown that *H. pylori* infection caused a marked induction of hBD-2. In contrast, in *H. pylori*-related gastritis, defensin gene expression is less pronounced [91].

It was also published that *H. pylori* stimulated the gastric epithelium to upregulate the endogenous production of AMPs (hBD-2). These authors also have shown that this process is regulated by the cytosolic pattern recognition receptor NOD1 [46]. A genetic study, analyzing single nucleotide polymorphisms, revealed that susceptibility to *H. pylori*-induced gastritis was due to constitutively expressed hBD1 [92].

It is known that chronic *H. pylori*-related gastritis causes replacement of normal mucosa by columnar epithelium with characteristics of the intestinal epithelium, called intestinal metaplasia, which occurs frequently [33]. In the condition mentioned earlier, Shen and coworkers observed a high HD-5 expression [34]. The authors suggested that Paneth cells present in the intestinal metaplasia could produce alpha defensins that enlarge the antibacterial response due to hBD-5 production. In addition, there is data confirming that *H. pylori* was able to induce cathelicidin LL-37 expression in gastric epithelial cells [52].

It has been estimated that about 10%–15% of the adult population in western countries suffers from cholelithiasis [93]. The majority of all gallstones remain asymptomatic; however infections of the biliary tree are still common and usually require antibiotic use. Under normal conditions bile is sterile; this effect is maintained due to the bactericidal effect of bile salts, immunoglobulins (IgA), and significant expression of hBD-1 and hBD-2 in the epithelium of the biliary tree and liver [94]. As observed in other parts of the gut, hBD-1 expression is constitutive; however hBD-2 could be stimulated in large intrahepatic bile ducts during obstruction or choledocholithiasis, where these peptides contribute to local antimicrobial defense [33]. On the other hand, in the epithelium of about 80% patients with primary sclerosing cholangitis (PSC), hBD-2 expression remained low [94]. Because in PSC patients, frequent bouts of serious infection are noticed. The observed lack of hBD-2 expression and possibly no other AMP induction, could explain the mechanism of the disease, as the authors suggest.

Bacterial colonization of the gut increases along the intestine; starting from the duodenum up to the proximal ileum, where the number of bacteria is relatively low. The distal part of the ileum contains about 10⁸ primarily anaerobic bacteria per gram of luminal contents [95], while there are up to 10^12 bacteria per gram in the colon. In this regard, the bacterial environment of the gut is crucial for the maintenance of human health and development of the mucosal immune system and, on the other hand, can contribute to the pathogenesis of chronic idiopathic bowel diseases [33].

Ulcerative colitis (UC) and Crohn's disease (CD) represent two principal manifestations of inflammatory bowel disease (IBD), a chronic inflammatory disorder of the digestive tract [96]. Both UC and CD present with similar symptoms including diarrhea, abdominal pain, weight loss, and fever, and both are characterized by a chronic and relapsing inflammation of the GI. In particular, CD is a chronic and segmental inflammation without a fully understood etiology; however, it results from the complex interaction between host genetics and the microbial population [97]. UC involves inflammation and ulceration of the colonic mucosa. The cause of both conditions is not a single organism, but a general microbial dysbiosis of the gut [98].

According to recent reports, the issue of expression of AMPs in IBD is controversial and needs to be thoroughly investigated. Interestingly, the level of AMPs is varied between the two diseases. Wehkamp et al. demonstrated that CD patients have an impaired Paneth cell production and maturation of hD-5 and hD-6 when compared to healthy subjects, and decreased expression of AMPs might be considered as the reason for bacterial induced mucosal inflammation and ileal involvement of CD [35]. However, it is questionable if the reduction in hD-5 expression is due to loss of the surface epithelium as a consequence of inflammatory changes rather that alterations in NOD2 status [37]. Mucosal hBD-1 expression is also decreased in both CD and UC [41]. In contrast, samples obtained from UC patients exhibited high mRNA levels of lysozyme, HD-5, HD-6, and hBD-2, which protects the mucosa due to both bactericidal and chemotactic properties of AMPs. However, their excessive production may contribute to inflammatory activity of UC, lymphocyte infiltration, and to the diarrhea affecting UC patients [59]. Expression of human cathelicidin is also increased in inflamed and non-inflamed mucosa in patients suffering from UC, but not in CD [48]. A similar phenomenon was observed when BPI (i.e., bactericidal/permeability-increasing protein) [99,100] and CCL20 [101] were investigated in the patients' colon tissues.

4 STIMULATED EXPRESSION OF AMPs IN THE DIGESTIVE TRACT AS A POTENTIAL NEW METHOD FOR INFECTION TREATMENT: A GREAT ALTERNATIVE TO ANTIBIOTIC THERAPY

A number of studies demonstrating the multifunctional roles of AMPs in the human GI tract suggest the clinical value of stimulating AMP expression and the possibility of using of these peptides for the treatment of pathogen-caused diseases in future applications. Previously, it was demonstrated that decreased expression of human cathelicidin during shigellosis might be restored after oral administration of sodium butyrate. Treatment of *Shigella*-infected rabbits with butyrate resulted in better clinical outcomes, reduced inflammation in the colon, and bacterial load in the stool, which was determined by upregulation of CAP-18 in the colonic epithelium and promotion of *Shigella* elimination [102]. In addition to that, it was presented that 4-phenylbutyrate (PBA), an odorless derivative of butyrate, enhances the expression of human cathelicidin in both epithelia of the colon and lung of rabbits, confirming the clinical usefulness of PBA in the treatment of *Shigella*-associated diarrhea and protection against secondary respiratory infections in the course of shigellosis [103]. On the other hand, both human and mouse cathelicidin were insufficient in the eradication of *Entamoeba histolytica* and in the treatment of colitis caused by this pathogen because *E. histolytica* degraded cathelicidin by released cysteine proteases [104]. Interestingly, intracolonic administration of the cathelicidin mCRAMP was used to treat mouse colitis, which allowed for the stimulation of MUC1 and MUC2/MAPK-dependent mucus synthesis and improvement of inflammatory parameters [105].

Importantly, results from a randomized double-blind placebo-controlled study clearly indicated the therapeutic value of vitamin D supplementation in the therapy of CD. Authors of the report demonstrated that short-term treatment with 2000 IU per day of vitamin D significantly recovers maintenance of small bowel and gastroduodenal permeability and increases the concentration of circulating LL-37, which results in the improvement of patient life [106]. The usefulness of vitamin D supplementation associated with increased LL-37 peptide level was also thoroughly tested in patients with new-onset severe sepsis or septic shock, but the results of this clinical trial need to be verified [107]. Due to impairment of maturation of HD-5 in CD patients, it was suggested that administration of exogenous mature HD-5 would be valuable in the treatment of this disease, as indicated using a DSS-treated mouse model [108].

5   CONCLUSIONS

Intestinal homeostasis is dependent on the firmly regulated dynamic between commensal bacteria, intestinal epithelial cells, and mucosal immune cells. Due to the constant contact of the GI tract with the external environment, AMPs play an emerging role in maintaining the integrity and balance of the GI tract. Recent studies clearly indicate that the knowledge about both the microbial-dependent and microbial-independent factors affecting AMP expression is crucial and might provide a novel therapeutic approach in the treatment of GI tract diseases associated with the colonization of tissues with pathogenic microbials and imbalance of host microbiota.

REFERENCES

[1] Wehkamp J, Schauber J, Stange EF. Defensins and cathelicidins in gastrointestinal infections. Curr Opin Gastroenterol 2007;23(1):32–8.

[2] Ohno H. Intestinal M cells. J Biochem 2016;159(2):151–60.

[3] Cabrera-Perez J, Badovinac VP, Griffith TS. Enteric immunity, the gut microbiome, and sepsis: rethinking the germ theory of disease. Exp Biol Med (Maywood) 2017;242(2):127–39.

[4] Filyk HA, Osborne LC. The multibiome: the intestinal ecosystem's influence on immune homeostasis, health, and disease. EBioMedicine 2016;13:46–54.

[5] Eckburg PB, et al. Diversity of the human intestinal microbial flora. Science 2005;308 (5728):1635–8.

[6] Koboziev I, et al. Role of the enteric microbiota in intestinal homeostasis and inflammation. Free Radic Biol Med 2014;68:122–33.

[7] Evans JM, Morris LS, Marchesi JR. The gut microbiome: the role of a virtual organ in the endocrinology of the host. J Endocrinol 2013;218(3):R37–47.

[8] Iebba V, et al. Eubiosis and dysbiosis: the two sides of the microbiota. New Microbiol 2016;39(1):1–12.

[9] Castanys-Muñoz E, Martin MJ, Vazquez E. Building a beneficial microbiome from birth. Adv Nutr 2016;7(2):323–30.

[10] Faith JJ, et al. The long-term stability of the human gut microbiota. Science 2013;341(6141).

[11] Giorgetti G, et al. Interactions between innate immunity, microbiota, and probiotics. J Immunol Res 2015;2015:501361.

[12] Kataoka K. The intestinal microbiota and its role in human health and disease. J Med Invest 2016;63(1–2):27–37.

[13] Walker A. Intestinal colonization and programming of the intestinal immune response. J Clin Gastroenterol 2014;48(Suppl 1):S8–11.

[14] Forbes JD, Van Domselaar G, Bernstein CN. The gut microbiota in immune-mediated inflammatory diseases. Front Microbiol 2016;7:1081.

[15] Dollé L, et al. Policing of gut microbiota by the adaptive immune system. BMC Med 2016;14:27.

[16] Fan D, et al. Activation of HIF-1α and LL-37 by commensal bacteria inhibits Candida albicans colonization. Nat Med 2015;21(7):808–14.

[17] Johansson ME, et al. The inner of the two Muc2 mucin-dependent mucus layers in colon is devoid of bacteria. Proc Natl Acad Sci U S A 2008;105(39):15064–9.

[18] Elia PP, et al. The role of innate immunity receptors in the pathogenesis of inflammatory bowel disease. Mediators Inflamm 2015;2015:936193.

[19] Gallo RL, Hooper LV. Epithelial antimicrobial defence of the skin and intestine. Nat Rev Immunol 2012;12(7):503–16.

[20] Mukherjee S, Hooper LV. Antimicrobial defense of the intestine. Immunity 2015;42 (1):28–39.

[21] Brugman S, et al. Mucosal immune development in early life: setting the stage. Arch Immunol Ther Exp (Warsz) 2015;63(4):251–68.

[22] Salzman NH, et al. Enteric defensins are essential regulators of intestinal microbial ecology. Nat Immunol 2010;11(1):76–83.

[23] Salzman NH, et al. Protection against enteric salmonellosis in transgenic mice expressing a human intestinal defensin. Nature 2003;422(6931):522–6.

[24] Frye M, et al. Differential expression of human alpha- and beta-defensins mRNA in gastrointestinal epithelia. Eur J Clin Invest 2000;30(8):695–701.

[25] Ghosh D, et al. Paneth cell trypsin is the processing enzyme for human defensin-5. Nat Immunol 2002;3(6):583–90.

[26] Nevalainen TJ, Grönroos JM, Kallajoki M. Expression of group II phospholipase A2 in the human gastrointestinal tract. Lab Invest 1995;72(2):201–8.

[27] Hooper LV, et al. Angiogenins: a new class of microbicidal proteins involved in innate immunity. Nat Immunol 2003;4(3):269–73.

[28] Cunliffe RN, Mahida YR. Expression and regulation of antimicrobial peptides in the gastrointestinal tract. J Leukoc Biol 2004;75(1):49–58.

[29] Kim JM. Antimicrobial proteins in intestine and inflammatory bowel diseases. Intest Res 2014;12(1):20–33.

[30] Zanetti M. The role of cathelicidins in the innate host defenses of mammals. Curr Issues Mol Biol 2005;7(2):179–96.

[31] Meyer-Hoffert U, et al. Secreted enteric antimicrobial activity localises to the mucus surface layer. Gut 2008;57(6):764–71.

[32] Wells JM, et al. Homeostasis of the gut barrier and potential biomarkers. Am J Physiol Gastrointest Liver Physiol 2017;312(3):G171–93.

[33] Jäger S, Stange EF, Wehkamp J. Antimicrobial peptides in gastrointestinal inflammation. Int J Inflam 2010;2010:910283.

[34] Shen B, et al. Human defensin 5 expression in intestinal metaplasia of the upper gastrointestinal tract. J Clin Pathol 2005;58(7):687–94.

[35] Wehkamp J, et al. NOD2 (CARD15) mutations in Crohn's disease are associated with diminished mucosal alpha-defensin expression. Gut 2004;53(11):1658–64.

[36] Ganz T. Defensins: antimicrobial peptides of innate immunity. Nat Rev Immunol 2003;3(9):710–20.

[37] Simms LA, et al. Reduced alpha-defensin expression is associated with inflammation and not NOD2 mutation status in ileal Crohn's disease. Gut 2008;57(7):903–10.

[38] Ryan LK, et al. Modulation of human beta-defensin-1 (hBD-1) in plasmacytoid dendritic cells (PDC), monocytes, and epithelial cells by influenza virus, Herpes simplex virus, and Sendai virus and its possible role in innate immunity. J Leukoc Biol 2011;90(2):343–56.

[39] Islam D, et al. Downregulation of bactericidal peptides in enteric infections: a novel immune escape mechanism with bacterial DNA as a potential regulator. Nat Med 2001;7(2):180–5.

[40] Sperandio B, et al. Virulent Shigella flexneri subverts the host innate immune response through manipulation of antimicrobial peptide gene expression. J Exp Med 2008;205(5):1121–32.

[41] Wehkamp J, et al. Inducible and constitutive beta-defensins are differentially expressed in Crohn's disease and ulcerative colitis. Inflamm Bowel Dis 2003;9(4):215–23.

[42] O'Neil DA, et al. Expression and regulation of the human beta-defensins hBD-1 and hBD-2 in intestinal epithelium. J Immunol 1999;163(12):6718–24.

[43] Schwab M, et al. The dietary histone deacetylase inhibitor sulforaphane induces human beta-defensin-2 in intestinal epithelial cells. Immunology 2008;125(2):241–51.

[44] Yin L, Chung WO. Epigenetic regulation of human β-defensin 2 and CC chemokine ligand 20 expression in gingival epithelial cells in response to oral bacteria. Mucosal Immunol 2011;4 (4):409–19.

[45] Kiehne K, et al. Oesophageal defensin expression during Candida infection and reflux disease. Scand J Gastroenterol 2005;40(5):501–7.

[46] Grubman A, et al. The innate immune molecule, NOD1, regulates direct killing of Helicobacter pylori by antimicrobial peptides. Cell Microbiol 2010;12(5):626–39.

[47] McDonald V, et al. A potential role for interleukin-18 in inhibition of the development of Cryptosporidium parvum. Clin Exp Immunol 2006;145(3):555–62.

[48] Schauber J, et al. Heterogeneous expression of human cathelicidin hCAP18/LL-37 in inflammatory bowel diseases. Eur J Gastroenterol Hepatol 2006;18(6):615–21.

[49] Yuk JM, et al. Vitamin D3 induces autophagy in human monocytes/macrophages via cathelicidin. Cell Host Microbe 2009;6(3):231–43.

[50] D'Aldebert E, et al. Bile salts control the antimicrobial peptide cathelicidin through nuclear receptors in the human biliary epithelium. Gastroenterology 2009;136(4):1435–43.

[51] Schauber J, et al. Control of the innate epithelial antimicrobial response is cell-type specific and dependent on relevant microenvironmental stimuli. Immunology 2006;118(4):509–19.

[52] Hase K, et al. Expression of LL-37 by human gastric epithelial cells as a potential host defense mechanism against Helicobacter pylori. Gastroenterology 2003;125(6):1613–25.

[53] Si-Tahar M, et al. Constitutive and regulated secretion of secretory leukocyte proteinase inhibitor by human intestinal epithelial cells. Gastroenterology 2000;118(6):1061–71.

[54] Nittayananta W, et al. Expression of oral secretory leukocyte protease inhibitor in HIV-infected subjects with long-term use of antiretroviral therapy. J Oral Pathol Med 2013;42 (3):208–15.

[55] Nicol AF, et al. Secretory leukocyte protease inhibitor expression and high-risk HPV infection in anal lesions of HIV-positive patients. J Acquir Immune Defic Syndr 2016;73(1):27–33.

[56] Hritz I, et al. Secretory leukocyte protease inhibitor expression in various types of gastritis: a specific role of Helicobacter pylori infection. Eur J Gastroenterol Hepatol 2006;18(3):277–82.

[57] Schmid M, et al. Attenuated induction of epithelial and leukocyte serine antiproteases elafin and secretory leukocyte protease inhibitor in Crohn's disease. J Leukoc Biol 2007;81 (4):907–15.

[58] Qu XD, et al. Secretion of type II phospholipase A2 and cryptdin by rat small intestinal Paneth cells. Infect Immun 1996;64(12):5161–5.

[59] Fahlgren A, et al. Increased expression of antimicrobial peptides and lysozyme in colonic epithelial cells of patients with ulcerative colitis. Clin Exp Immunol 2003;131(1):90–101.

[60] Salzman NH, et al. Enteric salmonella infection inhibits Paneth cell antimicrobial peptide expression. Infect Immun 2003;71(3):1109–15.

[61] Canny G, et al. Lipid mediator-induced expression of bactericidal/ permeability-increasing protein (BPI) in human mucosal epithelia. Proc Natl Acad Sci U S A 2002;99(6):3902–7.

[62] Balakrishnan A, Chakravortty D. Epithelial cell damage activates bactericidal/permeability increasing-protein (BPI) expression in intestinal epithelium. Front Microbiol 2017;8:1567.

[63] Santaolalla R, Abreu MT. Innate immunity in the small intestine. Curr Opin Gastroenterol 2012;28(2):124–9.

[64] Ogushi K, et al. Salmonella enteritidis FliC (flagella filament protein) induces human beta-defensin-2 mRNA production by Caco-2 cells. J Biol Chem 2001;276(32):30521–6.

[65] Vora P, et al. Beta-defensin-2 expression is regulated by TLR signaling in intestinal epithelial cells. J Immunol 2004;173(9):5398–405.

[66] Guo B, Cheng G. Modulation of the interferon antiviral response by the TBK1/IKKi adaptor protein TANK. J Biol Chem 2007;282(16):11817–26.

[67] Sotolongo J, et al. Host innate recognition of an intestinal bacterial pathogen induces TRIF-dependent protective immunity. J Exp Med 2011;208(13):2705–16.

[68] Wada A, et al. Induction of human beta-defensin-2 mRNA expression by Helicobacter pylori in human gastric cell line MKN45 cells on cag pathogenicity island. Biochem Biophys Res Commun 1999;263(3):770–4.

[69] Wehkamp J, et al. NF-kappaB- and AP-1-mediated induction of human beta defensin-2 in intestinal epithelial cells by Escherichia coli Nissle 1917: a novel effect of a probiotic bacterium. Infect Immun 2004;72(10):5750–8.

[70] Bu HF, et al. Lysozyme-modified probiotic components protect rats against polymicrobial sepsis: role of macrophages and cathelicidin-related innate immunity. J Immunol 2006;177 (12):8767–76.

[71] Girardin SE, et al. Nod1 detects a unique muropeptide from gram-negative bacterial peptidoglycan. Science 2003;300(5625):1584–7.

[72] Kim JG, Lee SJ, Kagnoff MF. Nod1 is an essential signal transducer in intestinal epithelial cells infected with bacteria that avoid recognition by toll-like receptors. Infect Immun 2004;72(3):1487–95.

[73] Rehman A, et al. Nod2 is essential for temporal development of intestinal microbial communities. Gut 2011;60(10):1354–62.

[74] Witthöft T, et al. Enhanced human beta-defensin-2 (hBD-2) expression by corticosteroids is independent of NF-kappaB in colonic epithelial cells (CaCo2). Dig Dis Sci 2005;50 (7):1252–9.

[75] Wang TT, et al. Cutting edge: 1,25-dihydroxyvitamin D3 is a direct inducer of antimicrobial peptide gene expression. J Immunol 2004;173(5):2909–12.

[76] Liu PT, et al. Convergence of IL-1beta and VDR activation pathways in human TLR2/1-induced antimicrobial responses. PLoS One 2009;4(6).

[77] Wang TT, et al. Direct and indirect induction by 1,25-dihydroxyvitamin D3 of the NOD2/CARD15-defensin beta2 innate immune pathway defective in Crohn disease. J Biol Chem 2010;285(4):2227–31.

[78] Makishima M, et al. Vitamin D receptor as an intestinal bile acid sensor. Science 2002;296 (5571):1313–6.

[79] Termén S, et al. PU.1 and bacterial metabolites regulate the human gene CAMP encoding antimicrobial peptide LL-37 in colon epithelial cells. Mol Immunol 2008;45 (15):3947–55.

[80] Gorr SU, Abdolhosseini M. Antimicrobial peptides and periodontal disease. J Clin Periodontol 2011;38(Suppl 11):126–41.

[81] Offenbacher S, et al. Gingival transcriptome patterns during induction and resolution of experimental gingivitis in humans. J Periodontol 2009;80(12):1963–82.

[82] Lu Q, et al. Expression of human beta-defensins-1 and -2 peptides in unresolved chronic periodontitis. J Periodontal Res 2004;39(4):221–7.

[83] Conti HR, et al. Th17 cells and IL-17 receptor signaling are essential for mucosal host defense against oral candidiasis. J Exp Med 2009;206(2):299–311.

[84] Hosokawa I, et al. Innate immune peptide LL-37 displays distinct expression pattern from beta-defensins in inflamed gingival tissue. Clin Exp Immunol 2006;146(2):218–25.

[85] Dale BA, Fredericks LP. Antimicrobial peptides in the oral environment: expression and function in health and disease. Curr Issues Mol Biol 2005;7(2):119–33.

[86] Hosaka Y, et al. Antimicrobial host defense in the upper gastrointestinal tract. Eur J Gastroenterol Hepatol 2008;20(12):1151–8.

[87] Steubesand N, et al. The expression of the beta-defensins hBD-2 and hBD-3 is differentially regulated by NF-kappaB and MAPK/AP-1 pathways in an in vitro model of Candida esophagitis. BMC Immunol 2009;10:36.

[88] Song JY, et al. The relationship between gastroesophageal reflux disease and chronic periodontitis. Gut Liver 2014;8(1):35–40.

[89] Marshall BJ, Warren JR. Unidentified curved bacilli in the stomach of patients with gastritis and peptic ulceration. Lancet 1984;1(8390):1311–5.

[90] Kusters JG, van Vliet AH, Kuipers EJ. Pathogenesis of Helicobacter pylori infection. Clin Microbiol Rev 2006;19(3):449–90.

[91] Wehkamp J, et al. Defensin pattern in chronic gastritis: HBD-2 is differentially expressed with respect to Helicobacter pylori status. J Clin Pathol 2003;56(5):352–7.

[92] Kocsis AK, et al. Potential role of human beta-defensin 1 in Helicobacter pylori-induced gastritis. Scand J Gastroenterol 2009;44(3):289–95.

[93] Stinton LM, Shaffer EA. Epidemiology of gallbladder disease: cholelithiasis and cancer. Gut Liver 2012;6(2):172–87.

[94] Harada K, et al. Peptide antibiotic human beta-defensin-1 and -2 contribute to antimicrobial defense of the intrahepatic biliary tree. Hepatology 2004;40(4):925–32.

[95] Berg RD. The indigenous gastrointestinal microflora. Trends Microbiol 1996;4(11):430–5.

[96] Yamada A, et al. Role of regulatory T cell in the pathogenesis of inflammatory bowel disease. World J Gastroenterol 2016;22(7):2195–205.

[97] Elson CO. Genes, microbes, and T cells—new therapeutic targets in Crohn's disease. N Engl J Med 2002;346(8):614–6.

[98] Lepage P, et al. Twin study indicates loss of interaction between microbiota and mucosa of patients with ulcerative colitis. Gastroenterology 2011;141(1):227–36.

[99] Haapamäki MM, et al. Bactericidal/permeability-increasing protein in colonic mucosa in ulcerative colitis. Hepatogastroenterology 1999;46(28):2273–7.

[100] Monajemi H, et al. Inflammatory bowel disease is associated with increased mucosal levels of bactericidal/permeability-increasing protein. Gastroenterology 1996;110(3):733–9.

[101] Kaser A, et al. Increased expression of CCL20 in human inflammatory bowel disease. J Clin Immunol 2004;24(1):74–85.

[102] Raqib R, et al. Improved outcome in shigellosis associated with butyrate induction of an endogenous peptide antibiotic. Proc Natl Acad Sci U S A 2006;103(24):9178–83.

[103] Sarker P, et al. Phenylbutyrate counteracts Shigella mediated downregulation of cathelicidin in rabbit lung and intestinal epithelia: a potential therapeutic strategy. PLoS One 2011;6(6).

[104] Cobo ER, et al. Entamoeba histolytica induces intestinal cathelicidins but is resistant to cathelicidin-mediated killing. Infect Immun 2012;80(1):143–9.

[105] Tai EK, et al. Cathelicidin stimulates colonic mucus synthesis by up-regulating MUC1 and MUC2 expression through a mitogen-activated protein kinase pathway. J Cell Biochem 2008;104(1):251–8.

[106] Raftery T, et al. Effects of vitamin D supplementation on intestinal permeability, cathelicidin and disease markers in Crohn's disease: results from a randomised double-blind placebo-controlled study. United European Gastroenterol J 2015;3(3):294–302.

[107] Quraishi SA, et al. Effect of cholecalciferol supplementation on vitamin D status and cathe-licidin levels in sepsis: a randomized, placebo-controlled trial. Crit Care Med 2015;43 (9):1928–37.

[108] Ishikawa C, et al. Precursor processing of human defensin-5 is essential to the multiple func-tions in vitro and in vivo. J Innate Immun 2010;2(1):66–76.

Chapter 2

Antimicrobial Peptides in the Host-Microbiota Homeostasis

Lin Zhang*,†,‡,a, Wei Hu*,a, Jeffery Ho*, Ross J. Fitzgerald§, Tony Gin*, Matthew T.V. Chan* and William K.K. Wu*,†

*Department of Anaesthesia and Intensive Care, The Chinese University of Hong Kong, Hong Kong SAR, China, †Institute of Digestive Diseases, State Key Laboratory of Digestive Diseases, LKS Institute of Health Sciences, CUHK Shenzhen Research Institute, The Chinese University of Hong Kong, Hong Kong SAR, China, ‡Laboratory of Molecular Pharmacology, Department of Pharmacology, School of Pharmacy, Southwest Medical University, Luzhou, Sichuan, China, §Roslin Institute, University of Edinburgh, Easter Bush Campus, Edinburgh, United Kingdom

Chapter Outline

1 BACKGROUND

Host-antimicrobial peptides (AMPs, also known as host-defense peptides) are the first line of defense against pathogens. Although in initially described as microbial agents, researchers have characterized them as modulators of inflammation and immunity [1]. These peptides are emerging as important in diseases affecting multiple organs including the lung [2–4], skin [5], and gastrointestinal (GI) tract [6–10]. Advances in sequencing technologies coupled with new bioinformatics developments have allowed the scientific community to begin to investigate the microbes that inhabit our body. However, the specific modulatory roles of AMPs in the host microbiota homeostasis remain under-researched

---

a. Equal contributors.

Antimicrobial Peptides in Gastrointestinal Diseases. https://doi.org/10.1016/B978-0-12-814319-3.00002-7
Copyright © 2018 Chi Hin Cho. Published by Elsevier Ltd. All rights reserved.
    21

and exploited. In this chapter, we focus on the connection between AMPs and host microbiota homeostasis. We first describe the composition of microbiota in human skin, lung, and gut, and then move to summarize AMPs function in maintaining/balancing microbiome homeostasis. Finally, we discuss future directions for studies of AMPs on microbiological disease disorders.

2 CHARACTERISTIC OF MICROBIOTA

In order to facilitate the characterization of human microbiota to further our knowledge of how the microbiome impacts human health and diseases, the National Institute of Health (NIH) Common Fund supported the Human Microbiome Project (HMP) in 2008 [11]. HMP1, the first phase of the HMP, has focused on producing reference genomes (viral, bacteria, and eukaryotic) with healthy individuals across different sites in the human body including: vaginal [12], oral cavity [13], skin [14], and GI tract [14–16]. The Integrative Human Microbiome Project (iHMP), in the second phase, started to create integrated longitudinal data sets of biological properties from both the microbiome and host from three different cohort studies of microbiome-associated conditions using multiple "omics" technologies [17]. Meanwhile, the European Commission financed MetaHIT in 2008 [18,19]. The central objective of MetaHIT was to establish associations between the genes of human intestinal microbiota and two disorders, namely inflammatory bowel disease (IBD) and obesity. With this great wave of microbiome studies, these multi-year and multicenter projects generated the largest and most comprehensive reference set of human microbiome [20].

Although the total bacterial load differs significantly between these body sites, such as skin, lung, and GI tract, the predominant bacteria phyla are very similar, consisting of *Firmicutes, Bacteriodetes, Proteobacteria, Actinobacteria, Fusobacteria,* and *Cyanobacteria* [21]. HMP has established a population-scale framework to develop metagenomics protocols [20]. The Microbiome Quality Control (MBQC) project baseline studies (MBQC-base) pave the way to guide researchers in experimental design choices for gut microbiome studies [22]. Although all habitats possess greater strain diversity, the baseline human microbial diversity has been finished with StrainPhlAn. This work enables an understanding of personalized microbiome function and dynamics [22].

2.1 Composition in the Skin Microbiome

Skin microbiota plays an intricate role in the human immune system—actually many immune functions—and helps to defend its host against invading bacterial pathogens [23,24]. With the sequencing technology advancement, the identified healthy human skin microbiota could be exploited in clinical diagnostic or therapeutic strategy. Resident microbiota may become pathogenic, sometimes in response to an impaired skin barrier [25]. This observation underscores the value of comprehensive characterization of the healthy skin microbiota to understand its role in the pathogenesis of skin disorders. With 16S rRNA sequencing technology, Gao et al. shed some light on the composition of superficial skin microbiota [26]. Further, HMP initiated a large-scale microbiome work and unveiled that the majority of bacterial division observed are *Proteobacteria* (16.5%), *Bacteroidetes* (6.3%), *Firmicutes* (24.4%), and *Actinobacteria* (51.8%) [27].

Many lines of evidence suggest a role for microorganisms even in noninfectious skin diseases, such as atopic dermatitis (AD), rosacea, psoriasis, and acne [28,29]. In these disease states, the microbiota was found in a dysbiosis condition. Advances in the next-generation sequencing have improved descriptions of the composition of the skin bacterial community known—as the microbiome—and furthered understanding of the degree of dysbiosis that is present on AD skin [30]. Both 16S and metagenomic analyses of bacterial DNAs have shown AD patients have an altered skin bacterial flora when compared to non-AD subjects [31,32]. Although it is not clear the underlying mechanism, people found the endogenous AMP levels decreased in the skin of subjects with AD [33]. AMP deficiency further results in altered balance in skin microbiota with overgrowth of pathogens such as *Staphylococcus aureus* (*S. aureus*) [34]. However, not only is *S. aureus* increased on AD skin but those specific strains of *S. aureus* are associated with more severe AD [35]. These observations, considered along with work that shows how normal commensal skin bacteria can promote health, support the hypothesis that dysbiosis in AD is critical in driving disease severity and/or tendency for relapse [30]. In a longitudinal cohort study, *S. aureus* was found to be more prevalent on the skin of infants who developed AD [36] Table 1.

Commensal bacteria colonized in normal skin promotes health by producing antimicrobial activity [44] by stimulation of AMPs or inhibition inflammation via cross talk with toll-like receptors [45]. Lai et al. found that normal bacteria living on the skin surface trigger a pathway that prevents excessive inflammation after injury. This study revealed a previously unknown mechanism by which a product from normal skin microflora could inhibit skin inflammation [45]. *S. epidermidis*, a major commensal bacterium, produces phenol-soluble modulins that inhibit pathogens such as *S. aureus* [46]. *Staphylococcal* lipoteichoic acid (LTA) inhibits *Propionibacterium acnes* (*P. acnes*) induced inflammation via the induction of miR-143. In this circumstance, it is possible that local regulation of the inflammatory responses by LTA at the site of acne vulgaris might therapeutically alleviate the *P. acnes*-induced inflammation [47]. Recent studies demonstrated that many constant assaults might harm the skin microbiota balance. These assaults include environmental agents, harsh cleansers and soaps, deodorants, and even medications and cosmetics [34].

TABLE 1 The Composition of Microbiota in the Skin, Lung and Gut

	Subjects	Microbiota Composition	References
Skin	Healthy	Actinobacteria, Firmicutes, Proteobacteria phylum	[26]
	Healthy	Proteobacteria, Bacteroidetes, Firmicutes, and Actinobacteria phylum	[27]
	Healthy	Proteobacteria, Bacteroidetes, Firmicutes, and Actinobacteria phylum	[30]
	Atopic dermatitis	Proteobacteria (decrease the relative abundance), Bacteroidetes (no change), Firmicutes (increase the absolute and relative abundance of *Staphylococcus*), and Actinobacteria (decrease the relative abundance)	[30]
	Atopic dermatitis	Increase the abundance of *Staphylococcus aureus* and skin commensal *Staphylococcus epidermidi, Staphylococcus hominis*	[31,34]
	Atopic dermatitis	Increase the abundance of *Staphylococcus aureus*	[32,34–36]
Lung	Healthy	Bacteroidetes, Firmicutes, Proteobacteria, and Actinobacteria phylum	[21]
	Healthy	*Prevotella, Streptococcus, Veilonella, Haemophilus, and Neisseria*	[37]
	Cystic fibrosis	Increase abundance of *Staphylococcus, Pseudomonas, or Achromobacter*	[37]
	Ventilator-associated pneumonia	Increase the *abundance of Burkholderia, Bacillales, Staphylococcus aureusa, and Pseudomonadales*	[38]
Gut	Heatlhy	Firmicutes and Bacteroidetes phylum	[39]
	Crohn's disease	Reduce the abundance of *Faecailbacterium prausnitzii* in Firmicutes phylum	[40]
	Inflammatory bowel disease	Increase the abundance of *Enterobacteriaceae* and *Escherichia coli* from Proteobacteria phylum	[40]
	Diabetes	Reduce the abundance of *Roseburia intestinalis* and *Faecalibacterium prausnitzi*	[41]

TABLE 1 The Composition of Microbiota in the Skin, Lung and Gut—cont'd

Subjects	Microbiota Composition	References
Hypertension	Increase the abundance of Firmicutes and Bacteroidetes phylum	[42]
Colon cancer	Increase the abundane of *Fusobacterium nucleatu, Peptostreptococcus anaerobius,* and *Parvimonas micra*	[43]

2.2   Composition in the Respiratory Tract Microbiome

In the last decade, the concept of "a healthy lung is a sterile lung" was challenged by use of high-throughput sequencing methods (next-generation sequencing, NGS). Now people realize that the respiratory microbiome might play a critical role in the development of lung disease [21,48]. The healthy lung microbiota is thought to be diverse, and a recent study indicated that the airway microbiota reaches a greater diversity more rapidly than the intestinal microbiota following birth. The most prevalent phyla in the airways are *Bacteroidetes* and *Firmicutes*, and to a lesser extent *Proteobacteria* and *Actinobacteria*. This core airway microbiota is mainly comprised of genera *Pseudomonas* (Protebacteria), *Streptococcus* (Firmicutes), *Prevotella* (Bacterioidetes), *Fusobacteria* (Fusobacteria), *Veillonella* (Firmicutes), *Haemophilus* (Proteobacteria), *Neisseria* (Proteobacteria), and *Porphyromonas* (Bacterioidetes) [21]. For lung pathogen microbiome, one study using WGS sequencing for cystic fibrosis demonstrated that sputum DNAs provide valuable information beyond the possibilities of culture-based diagnosis. *Prevotella, Streptococcus, Veilonella, Haemophilus,* and *Neisseria* are the most abundant taxa in healthy subjects, while *Pseudomonas, Staphylococcus, Stenotrophomonas,* or *Achromobacter* often harbored in the lung of cystic fibrosis patients. The further functional and taxonomic features of dominant pathogens, including antibiotic resistances, can be deduced, supporting accurate and non-invasive clinical diagnosis [37].

Dysbiosis of the respiratory microbiome is more profound in patients who develop ventilator-associated pneumonia (VAP) than in those that do not develop pneumonia in the Intensive Care Unit (ICU). *Burkholderia, Bacillales, S. aureus,* and *Pseudomonadales* are found to be related positively associated during mechanical ventilation [38]. The healthy and respiratory inflamed microbiota from claves upper and lower respiratory tracts were also characterized using 16S rRNA gene sequencing. Results suggest that the composition of the upper airway microbiota in cattle was affected by environmental factors and it influenced the lung microbiota as well [49]. Respiratory research has entered a new era in which host-microorganism interactions need to be considered (Table 1).

2.3   Composition in the Gastrointestinal Tract Microbiome

The mammalian GI tract is host to trillions of microorganisms (including bacteria, fungi, and viruses). In the past two decades, our knowledge of the composition of gut microbiome and their potential functional role is booming. Gut microbiota has been recognized as a fundamental player in almost all diseases, colon cancer [43], cancer therapy [39], diabetes [41], IBD [40], infection [50–52], hypertension [42], and others (Table 1). Commensal bacteria protect from pathogen invasion, extract additional energy from food, and synthesize key molecules for tissue development in a way that is highly specialized with respect to their location along the GI tract [53]. Antibiotics are invaluable weapons to fight infectious diseases. However, by altering the composition and functions of the microbiota, they can also produce long-lasting deleterious effects for the host [54]. Recent investigations suggest that the efficacy of some clinical approaches depends on the action of commensal bacteria [55]. Commensal anaerobic bacteria in the gut provide a key defense mechanism by inhibiting the growth of potentially pathogenic bacteria [56]. One possible mechanism for maintaining pathogenic bacteria GI colonization resistance involves commensal anaerobe induction of mucosal immune effectors (e.g., antimicrobial peptides, AMPs) that kill the pathogen [57].

3   BIOLOGICAL FUNCTION OF ANTIMICROBIAL PEPTIDES

Endogenous antimicrobial peptides (AMPs), such as cathelicidins [58,59] and defensins [33] are among the most ancient and efficient components of host defense systems. These AMPs very often possess broad-spectrum microbicidal activity against a wide range of microbes, including Gram-positive [34], Gram-negative bacteria [7,9], and fungi [60]. Dependent on the presence of α-helical or β-sheet structures, AMPs are classified into four groups: α-, β-, αβ-, and non-αβ families [61]. These peptides provide non-specific innate immune defense against microbes via pore formation on bacterial membranes, regulate inflammatory response through neutralization of lipopolysaccharides, modulation of cytokine expression, and promote tissue repair via increasing angiogenesis and growth signals [1,62]. The importance of AMPs and the intricate regulation of their bactericidal activity could be exemplified by the increased susceptibility to invasive *Group A Streptococcus* [59] and *Staphylococcus aureus* [63] skin infections in cathelicidin-knockout mice, and the redox-activation of human β-defensin 1 against colonization by commensal bacteria and opportunistic fungi [64].

Our research team has a longstanding track record in the functional investigation of AMPs, in particular cathelicidin. We are among the first to demonstrate that cathelicidin could protect against *Helicobacter pylori* infection [7,9], inhibit GI tumorigenesis [65,66], promote tissue repair [67,68], and modulate inflammation [67]. Importantly, for improving the efficacy of *in-vivo* delivery of cathelicidin, our team successfully bioengineered a probiotic strain of lactic acid bacteria that could mediate the mucosal release of this peptide [6,7,9,10]. In an experimental model of colitis, our team also discovered that cathelicidin could preserve the mucosal integrity by promoting mucus secretion through upregulating mucin gene expression and reducing apoptosis of intestinal epithelial cells [10]. Our recent systematic review also reported that AMPs could regulate immune response, pyroptosis, and coagulation cascades in the protection against sepsis (a systemic host response to an infection leading to organ failure) [69].

4   ANTIMICROBIAL PEPTIDE AS A TARGETED MODULATOR ON HUMAN MICROBIOME

AMPs belong to an ancient defense system found in all organisms and participate in a preservative co-evolution with a complex microbiome [70]. Emerging evidence points to a possible central role for AMPs in determining the composition of these microbes. Richard Gallo's research team first reported that skin microbiome provides the first line of defense, whereas the host innate immune system provides the second line of defense, which is activated only after the surface is damaged to trigger host AMP production [34]. AMPs from *S. epidermidis* were also known to exert selective killing, a logical behavior if the cell is to resist killing itself [29] and the capacity for selective killing of pathogenic bacteria over normal microflora is highly desirable because it will help to shape the normal bacterial community [34].

With over two decades of exploration on host microbiome, however, it is still lacking effective therapeutic targets on specific individual species. AMPs can be secreted by intestinal epithelial and protect the host against bacteria, fungi, and some viruses and shape the composition of the intestinal microbiota [71]. Mice studies have shown that the expression level of AMPs of the α-defensin family greatly affects community composition [72]. In researching whether AMPs could target pathogens other than commensal in the gut, Cullen et al. found that human gut commensal microbes are resistant to high levels of inflammation-associated AMPs. They also identified a mechanism for AMPs' resistance due to lipopolysaccharide (LPS) modification in the phylum *Bacteroidetes*. In certain pathogens, however, lipid A modification may provide additional benefits to commensal microbes beyond AMP resistance [73]. Specific AMPs account for different bacterial communities associated with closely related species of the cnidarian *Hydra*. shape species-specific bacterial communities [74]. Interestingly, an antimicrobial peptide named C16G2 was able to selectively target killing cariogenic pathogen *Streptococcus mutans in vitro* oral multispecies community [75]. Teshima et al. discovered that the molecule R-Spondin 1 stimulates intestinal stem cells to differentiate Paneth cells to secrete multiple AMPs, which has strong and selective antimicrobial activities against pathological bacteria and does not kill symbiotic microbes in a healthy gut. Thus, the administration of R-Spo1 represents a novel and physiological approach to restore the gut's ecosystem and homeostasis while avoiding adverse effects [76].

5 CHALLENGES AND FUTURE AVENUES

In this chapter, we have summarized AMPs on human microbiota homeostasis, providing an overview of human skin, lung, and gut microbiome composition, updating our knowledge on how AMPs modulate microbiota, and discussed the therapeutic implication of AMPs for different disorders. The next challenge will be how to modify the patient's microbiota by AMPs properly. In better understanding the mechanisms of commensal and pathogenic bacteria adaption to AMPs we expect to open new AMPs-based therapeutic avenues to conquer disease due to microbiota dysbiosis. Commensal bacterial development of AMPs resistance may be a crucial mechanism of host microbiota homeostasis. Given that AMPs are critical to limit pathogens colonization, it is tempting to postulate that development of resistance to AMPs might support host microbiome dysbiosis. Therefore, it would be interesting to elucidate the genetic mechanism's underlying acquisition of resistance to human AMPs by specific commensal or pathogen strains. Although the field of therapeutic intervention through targeting microbiota is still primitive, some approaches are already possible. Further detailed metagenomics, transcriptomic, proteomic, and metabolomics analyses will provide new insights into the role of AMPs in host microbiota homeostasis and in the composition of microbiota during health and diseases.

ACKNOWLEDGMENTS

This work was supported by the Health and Medical Research Fund (16151172, to LZ; and 16151002, to WKKW) and the National Natural Science Foundation of China (81402014, to LZ; and 81770562, to CHC).

REFERENCES

[1] Chow JY, Li ZJ, Wu WK, Cho CH. Cathelicidin a potential therapeutic peptide for gastrointestinal inflammation and cancer. World J Gastroenterol 2013;19(18):2731–5.

[2] Li D, Beisswenger C, Herr C, Schmid RM, Gallo RL, Han G, Zakharkina T, Bals R. Expression of the antimicrobial peptide cathelicidin in myeloid cells is required for lung tumor growth. Oncogene 2014;33(21):2709–16.

[3] Kovach MA, Ballinger MN, Newstead MW, Zeng X, Bhan U, Yu FS, Moore BB, Gallo RL, Standiford TJ. Cathelicidin-related antimicrobial peptide is required for effective lung mucosal immunity in gram-negative bacterial pneumonia. J Immunol 2012;189(1):304–11.

[4] Byfield FJ, Kowalski M, Cruz K, Leszczynska K, Namiot A, Savage PB, Bucki R, Janmey PA. Cathelicidin ll-37 increases lung epithelial cell stiffness, decreases transepithelial permeability, and prevents epithelial invasion by pseudomonas aeruginosa. J Immunol 2011;187(12):6402–9.

[5] Zhang LJ, Sen GL, Ward NL, Johnston A, Chun K, Chen Y, Adase C, Sanford JA, Gao N, Chensee M, Sato E, et al. Antimicrobial peptide ll37 and mavs signaling drive interferon-beta production by epidermal keratinocytes during skin injury. Immunity 2016;45(1):119–30.

[6] Wong CC, Zhang L, Wu WK, Shen J, Chan RL, Lu L, Hu W, Li MX, Li LF, Ren SX, Li YF, et al. Cathelicidin-encoding lactococcus lactis promotes mucosal repair in murine experimental colitis. J Gastroenterol Hepatol 2017;32(3):609–19.

[7] Zhang L, Wu WK, Gallo RL, Fang EF, Hu W, Ling TK, Shen J, Chan RL, Lu L, Luo XM, Li MX, et al. Critical role of antimicrobial peptide cathelicidin for controlling helicobacter pylori survival and infection. J Immunol 2016;196(4):1799–809.

[8] Wu WK, Sung JJ, Cheng AS, Chan FK, Ng SS, To KF, Wang XJ, Zhang L, Wong SH, Yu J, Cho CH. The janus face of cathelicidin in tumorigenesis. Curr Med Chem 2014;21(21):2392–400.

[9] Zhang L, Yu J, Wong CC, Ling TK, Li ZJ, Chan KM, Ren SX, Shen J, Chan RL, Lee CC, Li MS, et al. Cathelicidin protects against helicobacter pylori colonization and the associated gastritis in mice. Gene Ther 2013;20(7):751–60.

[10] Wong CC, Zhang L, Li ZJ, Wu WK, Ren SX, Chen YC, Ng TB, Cho CH. Protective effects of cathelicidin-encoding lactococcus lactis in murine ulcerative colitis. J Gastroenterol Hepatol 2012;27(7):1205–12.

[11] Human Microbiome Project Consortium. Structure, function and diversity of the healthy human microbiome. Nature 2012;486(7402):207–14.

[12] Si J, You HJ, Yu J, Sung J, Ko G. Prevotella as a hub for vaginal microbiota under the influence of host genetics and their association with obesity. Cell Host Microbe 2017;21(1):97–105.

[13] Gonzalez A, Hyde E, Sangwan N, Gilbert JA, Viirre E, Knight R. Migraines are correlated with higher levels of nitrate-, nitrite-, and nitric oxide-reducing oral microbes in the american gut project cohort. mSystems 2016;1(5):e00105–16.

[14] Zhernakova A, Kurilshikov A, Bonder MJ, Tigchelaar EF, Schirmer M, Vatanen T, Mujagic Z, Vila AV, Falony G, Vieira-Silva S, Wang J, et al. Population-based metagenomics analysis reveals markers for gut microbiome composition and diversity. Science 2016; 352(6285):565–9.

[15] Beaumont M, Goodrich JK, Jackson MA, Yet I, Davenport ER, Vieira-Silva S, Debelius J, Pallister T, Mangino M, Raes J, Knight R, et al. Heritable components of the human fecal microbiome are associated with visceral fat. Genome Biol 2016;17(1):189.

[16] Falony G, Joossens M, Vieira-Silva S, Wang J, Darzi Y, Faust K, Kurilshikov A, Bonder MJ, Valles-Colomer M, Vandeputte D, Tito RY, et al. Population-level analysis of gut microbiome variation. Science 2016;352(6285):560–4.

[17] Integrative HMPRNC. The integrative human microbiome project: dynamic analysis of microbiome-host omics profiles during periods of human health and disease. Cell Host Microbe 2014;16(3):276–89.

[18] Qin J, Li R, Raes J, Arumugam M, Burgdorf KS, Manichanh C, Nielsen T, Pons N, Levenez F, Yamada T, Mende DR, et al. A human gut microbial gene catalogue established by metagenomic sequencing. Nature 2010;464(7285):59–65.

[19] Li J, Jia H, Cai X, Zhong H, Feng Q, Sunagawa S, Arumugam M, Kultima JR, Prifti E, Nielsen T, Juncker AS, et al. An integrated catalog of reference genes in the human gut microbiome. Nat Biotechnol 2014;32(8):834–41.

[20] Human Microbiome Project Consortium. A framework for human microbiome research. Nature 2012;486(7402):215–21.

[21] Marsland BJ, Gollwitzer ES. Host-microorganism interactions in lung diseases. Nat Rev Immunol 2014;14(12):827–35.

[22] Sinha R, Abu-Ali G, Vogtmann E, Fodor AA, Ren B, Amir A, Schwager E, Crabtree J, Ma S. Microbiome Quality Control Project C, Abnet CC et al: Assessment of variation in microbial community amplicon sequencing by the microbiome quality control (mbqc) project consortium. Nat Biotechnol (2017) 35(11):1077–1086.

[23] Oh J, Byrd AL, Park M, Program NCS, Kong HH, Segre JA. Temporal stability of the human skin microbiome. Cell 2016;165(4):854–66.

[24] Dorrestein PC, Gallo RL, Knight R. Microbial skin inhabitants: friends forever. Cell 2016; 165(4):771–2.

[25] Musharrafieh R, Tacchi L, Trujeque J, LaPatra S, Salinas I. Staphylococcus warneri, a resident skin commensal of rainbow trout (Oncorhynchus mykiss) with pathobiont characteristics. Veterinary Microbiology 2014;169(1–2):80–8.

[26] Gao Z, Tseng CH, Pei Z, Blaser MJ. Molecular analysis of human forearm superficial skin bacterial biota. Proc Natl Acad Sci U S A 2007;104(8):2927–32.

[27] Grice EA, Kong HH, Renaud G, Young AC, Program NCS, Bouffard GG, Blakesley RW, Wolfsberg TG, Turner ML, Segre JA. A diversity profile of the human skin microbiota. Genome Res 2008;18(7):1043–50.

[28] Thomsen RJ, Stranieri A, Knutson D, Strauss JS. Topical clindamycin treatment of acne. Clinical, surface lipid composition, and quantitative surface microbiology response. Arch Dermatol 1980;116(9):1031–4.

[29] Paulino LC, Tseng CH, Strober BE, Blaser MJ. Molecular analysis of fungal microbiota in samples from healthy human skin and psoriatic lesions. J Clin Microbiol 2006;44(8):2933–41.

[30] Williams MR, Gallo RL. The role of the skin microbiome in atopic dermatitis. Curr Allergy Asthma Rep 2015;15(11):65.

[31] Kong HH, Oh J, Deming C, Conlan S, Grice EA, Beatson MA, Nomicos E, Polley EC, Komarow HD, Program NCS, Murray PR, et al. Temporal shifts in the skin microbiome associated with disease flares and treatment in children with atopic dermatitis. Genome Res 2012;22(5):850–9.

[32] Leyden JJ, Marples RR, Kligman AM. Staphylococcus aureus in the lesions of atopic dermatitis. Br J Dermatol 1974;90(5):525–30.

[33] Ong PY, Ohtake T, Brandt C, Strickland I, Boguniewicz M, Ganz T, Gallo RL, Leung DY. Endogenous antimicrobial peptides and skin infections in atopic dermatitis. N Engl J Med 2002;347(15):1151–60.

[34] Nakatsuji T, Chen TH, Narala S, Chun KA, Two AM, Yun T, Shafiq F, Kotol PF, Bouslimani A, Melnik AV, Latif H, et al. Antimicrobials from human skin commensal bacteria protect against Staphylococcus aureus and are deficient in atopic dermatitis. Sci Transl Med 2017;9(378).

[35] Byrd AL, Deming C, Cassidy SKB, Harrison OJ, Ng WI, Conlan S, Program NCS, Belkaid Y, Segre JA, Kong HH. Staphylococcus aureus and Staphylococcus epidermidis strain diversity underlying pediatric atopic dermatitis. Sci Transl Med 2017;9(397):eaal4651.

[36] Meylan P, Lang C, Mermoud S, Johannsen A, Norrenberg S, Hohl D, Vial Y, Prod'hom G, Greub G, Kypriotou M, Christen-Zaech S. Skin colonization by Staphylococcus aureus precedes the clinical diagnosis of atopic dermatitis in infancy. J Invest Dermatol 2017; 137(12):2497–504.

[37] Feigelman R, Kahlert CR, Baty F, Rassouli F, Kleiner RL, Kohler P, Brutsche MH, von Mering C, Sputum DNA. sequencing in cystic fibrosis: non-invasive access to the lung microbiome and to pathogen details. Microbiome 2017;5(1):20.

[38] Zakharkina T, Martin-Loeches I, Matamoros S, Povoa P, Torres A, Kastelijn JB, Hofstra JJ, de Wever B, de Jong M, Schultz MJ, Sterk PJ, et al. The dynamics of the pulmonary microbiome

during mechanical ventilation in the intensive care unit and the association with occurrence of pneumonia. Thorax 2017;72(9):803–10.

[39] Roy S, Trinchieri G. Microbiota: a key orchestrator of cancer therapy. Nat Rev Cancer 2017; 17(5):271–85.

[40] Ni J, Wu GD, Albenberg L, Tomov VT: Gut microbiota and ibd: causation or correlation? Nature Reviews Gastroenterology & Hepatology (2017) 14(10):573–584.

[41] Tilg H, Moschen AR. Microbiota and diabetes: an evolving relationship. Gut 2014; 63(9):1513–21.

[42] Jose PA, Raj D. Gut microbiota in hypertension. Curr Opin Nephrol Hypertens 2015; 24(5):403–9.

[43] Wong S, Kwong T, Chow T, Luk A, Dai R, Nakatsu G, Lam T, Zhang L, Wu J, Chan F, Ng S, Wong M, Ng S, Wu W, Yu J, Sung J. Quantitation of faecal Fusobacterium improves faecal immunochemical test in detecting advanced colorectal neoplasia. Gut 2017;66(8):1441–8.

[44] Frick IM, Nordin SL, Baumgarten M, Morgelin M, Sorensen OE, Olin AI, Egesten A. Constitutive and inflammation-dependent antimicrobial peptides produced by epithelium are differentially processed and inactivated by the commensal finegoldia magna and the pathogen streptococcus pyogenes. J Immunol 2011;187(8):4300–9.

[45] Lai Y, Di Nardo A, Nakatsuji T, Leichtle A, Yang Y, Cogen AL, Wu ZR, Hooper LV, Schmidt RR, von Aulock S, Radek KA, et al. Commensal bacteria regulate toll-like receptor 3-dependent inflammation after skin injury. Nat Med 2009;15(12):1377–82.

[46] Cogen AL, Yamasaki K, Sanchez KM, Dorschner RA, Lai Y, MacLeod DT, Torpey JW, Otto M, Nizet V, Kim JE, Gallo RL. Selective antimicrobial action is provided by phenol-soluble modulins derived from Staphylococcus epidermidis, a normal resident of the skin. J Invest Dermatol 2010;130(1):192–200.

[47] Xia X, Li Z, Liu K, Wu Y, Jiang D, Lai Y. Staphylococcal lta-induced mir-143 inhibits propionibacterium acnes-mediated inflammatory response in skin. J Invest Dermatol 2016;136(3):621–30.

[48] Tracy M, Cogen J, Hoffman LR. The pediatric microbiome and the lung. Curr Opin Pediatr 2015;27(3):348–55.

[49] Nicola I, Cerutti F, Grego E, Bertone I, Gianella P, D'Angelo A, Peletto S, Bellino C: Characterization of the upper and lower respiratory tract microbiota in piedmontese calves. Microbiome (2017) 5(1):152.

[50] McKenney PT, Pamer EG. From hype to hope: the gut microbiota in enteric infectious disease. Cell 2015;163(6):1326–32.

[51] Kane M, Case LK, Kopaskie K, Kozlova A, MacDearmid C, Chervonsky AV, Golovkina TV. Successful transmission of a retrovirus depends on the commensal microbiota. Science 2011;334(6053):245–9.

[52] Kuss SK, Best GT, Etheredge CA, Pruijssers AJ, Frierson JM, Hooper LV, Dermody TS, Pfeiffer JK. Intestinal microbiota promote enteric virus replication and systemic pathogenesis. Science 2011;334(6053):249–52.

[53] Brestoff JR, Artis D. Commensal bacteria at the interface of host metabolism and the immune system. Nat Immunol 2013;14(7):676–84.

[54] Becattini S, Taur Y, Pamer EG. Antibiotic-induced changes in the intestinal microbiota and disease. Trends Mol Med 2016;22(6):458–78.

[55] Routy B, Le Chatelier E, Derosa L, Duong CPM, Alou MT, Daillere R, Fluckiger A, Messaoudene M, Rauber C, Roberti MP, Fidelle M, et al. Gut microbiome influences efficacy of pd-1-based immunotherapy against epithelial tumors. Science 2018;359(6371):91–7.

[56] Ubeda C, Djukovic A, Isaac S. Roles of the intestinal microbiota in pathogen protection. Clinical & translational immunology 2017;6(2).

[57] Ostaff MJ, Stange EF, Wehkamp J. Antimicrobial peptides and gut microbiota in homeostasis and pathology. EMBO Mol Med 2013;5(10):1465–83.

[58] Chromek M, Slamova Z, Bergman P, Kovacs L, Podracka L, Ehren I, Hokfelt T, Gudmundsson GH, Gallo RL, Agerberth B, Brauner A. The antimicrobial peptide cathelicidin protects the urinary tract against invasive bacterial infection. Nat Med 2006;12(6):636–41.

[59] Nizet V, Ohtake T, Lauth X, Trowbridge J, Rudisill J, Dorschner RA, Pestonjamasp V, Piraino J, Huttner K, Gallo RL. Innate antimicrobial peptide protects the skin from invasive bacterial infection. Nature 2001;414(6862):454–7.

[60] Lopez-Garcia B, Lee PH, Yamasaki K, Gallo RL. Anti-fungal activity of cathelicidins and their potential role in Candida albicans skin infection. J Invest Dermatol 2005;125(1):108–15.

[61] Wu WK, Wang G, Coffelt SB, Betancourt AM, Lee CW, Fan D, Wu K, Yu J, Sung JJ, Cho CH. Emerging roles of the host defense peptide ll-37 in human cancer and its potential therapeutic applications. Int J Cancer 2010;127(8):1741–7.

[62] Wu WK, Wong CC, Li ZJ, Zhang L, Ren SX, Cho CH. Cathelicidins in inflammation and tissue repair: potential therapeutic applications for gastrointestinal disorders. Acta Pharmacol Sin 2010;31(9):1118–22.

[63] Zhang LJ, Guerrero-Juarez CF, Hata T, Bapat SP, Ramos R, Plikus MV, Gallo RL. Innate immunity. Dermal adipocytes protect against invasive Staphylococcus aureus skin infection. Science 2015;347(6217):67–71.

[64] Schroeder BO, Wu Z, Nuding S, Groscurth S, Marcinowski M, Beisner J, Buchner J, Schaller M, Stange EF, Wehkamp J. Reduction of disulphide bonds unmasks potent antimicrobial activity of human beta-defensin 1. Nature 2011;469(7330):419–23.

[65] Ren SX, Cheng AS, To KF, Tong JH, Li MS, Shen J, Wong CC, Zhang L, Chan RL, Wang XJ, Ng SS, et al. Host immune defense peptide ll-37 activates caspase-independent apoptosis and suppresses colon cancer. Cancer Res 2012;72(24):6512–23.

[66] Wu WK, Sung JJ, To KF, Yu L, Li HT, Li ZJ, Chu KM, Yu J, Cho CH. The host defense peptide ll-37 activates the tumor-suppressing bone morphogenetic protein signaling via inhibition of proteasome in gastric cancer cells. J Cell Physiol 2010;223(1):178–86.

[67] Tai EK, Wu WK, Wang XJ, Wong HP, Yu L, Li ZJ, Lee CW, Wong CC, Yu J, Sung JJ, Gallo RL, et al. Intrarectal administration of mcramp-encoding plasmid reverses exacerbated colitis in cnlp(-/-) mice. Gene Ther 2013;20(2):187–93.

[68] Yang YH, Wu WK, Tai EK, Wong HP, Lam EK, So WH, Shin VY, Cho CH. The cationic host defense peptide rcramp promotes gastric ulcer healing in rats. J Pharmacol Exp Ther 2006;318 (2):547–54.

[69] Ho J, Zhang L, Liu X, Wong SH, Wang MHT, Lau BWM, Ngai SPC, Chan H, Choi G, Leung CCH, Wong WT, et al. Pathological role and diagnostic value of endogenous host defense peptides in adult and neonatal sepsis: a systematic review. Shock 2017;47(6):673–9.

[70] Foster KR, Schluter J, Coyte KZ, Rakoff-Nahoum S. The evolution of the host microbiome as an ecosystem on a leash. Nature 2017;548(7665):43–51.

[71] Schroeder BO, Ehmann D, Precht JC, Castillo PA, Kuchler R, Berger J, Schaller M, Stange EF, Wehkamp J. Paneth cell alpha-defensin 6 (hd-6) is an antimicrobial peptide. Mucosal Immunol 2015;8(3):661–71.

[72] Salzman NH, Hung K, Haribhai D, Chu H, Karlsson-Sjoberg J, Amir E, Teggatz P, Barman M, Hayward M, Eastwood D, Stoel M, et al. Enteric defensins are essential regulators of intestinal microbial ecology. Nat Immunol 2010;11(1):76–83.

[73] Cullen TW, Schofield WB, Barry NA, Putnam EE, Rundell EA, Trent MS, Degnan PH, Booth CJ, Yu H, Goodman AL. Gut microbiota. Antimicrobial peptide resistance mediates resilience of prominent gut commensals during inflammation. Science 2015;347(6218):170–5.

[74] Franzenburg S, Walter J, Kunzel S, Wang J, Baines JF, Bosch TC, Fraune S. Distinct antimicrobial peptide expression determines host species-specific bacterial associations. Proc Natl Acad Sci U S A 2013;110(39):E3730–3738.

[75] Guo L, McLean JS, Yang Y, Eckert R, Kaplan CW, Kyme P, Sheikh O, Varnum B, Lux R, Shi W, He X. Precision-guided antimicrobial peptide as a targeted modulator of human microbial ecology. Proc Natl Acad Sci U S A 2015;112(24):7569–74.

[76] Hayase E, Hashimoto D, Nakamura K, Noizat C, Ogasawara R, Takahashi S, Ohigashi H, Yokoi Y, Sugimoto R, Matsuoka S, Ara T, et al. R-spondin1 expands paneth cells and prevents dysbiosis induced by graft-versus-host disease. J Exp Med 2017;214(12):3507–18.

FURTHER READING

Vujkovic-Cvijin I, Dunham RM, Iwai S, Maher MC, Albright RG, Broadhurst MJ, Hernandez RD, Lederman MM, Huang Y, Somsouk M, Deeks SG, et al. Dysbiosis of the gut microbiota is associated with hiv disease progression and tryptophan catabolism. Sci Transl Med 2013;5(193).

Chapter 3

The Roles of Antimicrobial Peptides in the Regulation of Gastrointestinal Microbiota and Innate Immunity

Ivy K.M. Law*, Michelle W. Cheng*, David Q. Shih‡, Dermot P.B. McGovern† and Hon Wai Koon*

*University of California Los Angeles, Los Angeles, CA, United States, †Inflammatory Bowel & Immunobiology Research Institute, Los Angeles, CA, United States, ‡Comprehensive Digestive Institute of Nevada, Las Vegas, NV, United States

Chapter Outline

1 A BRIEF INTRODUCTION TO GASTROINTESTINAL ANTIMICROBIAL PEPTIDES

Antimicrobial peptides (AMPs) are important modulators of gastrointestinal microbiota and innate immunity [1]. Their expression in the gut is dependent on intestinal infection and inflammation. Gastrointestinal AMPs include alpha-defensin, beta-defensin, cathelicidin, elafin, SLPI, lipocalin, calprotectin, lactoferrin, and others [1]. Some of the AMPs are constitutively expressed while others are induced by pathogen-associated molecular patterns (PAMPs). Intestinal epithelial cells, immune cells, and Paneth cells are main sources of gastrointestinal AMPs.

Antimicrobial Peptides in Gastrointestinal Diseases. https://doi.org/10.1016/B978-0-12-814319-3.00003-9
Copyright © 2018 Chi Hin Cho. Published by Elsevier Ltd. All rights reserved.

Humans and mice have only one form of cathelicidin, called LL-37 and mCRAMP, respectively [1,2]. Cathelicidin exhibits antimicrobial effects against group A *Streptococcus*, *Staphylococcus aureus*, enteroinvasive *Escherichia coli* O29 [3] and also *Helicobacter pylori* [4]. LL-37 can form transmembrane pores on the cell membrane of target organisms [5]. As a result, cathelicidin increases cell membrane permeability and inhibits bacterial cell wall synthesis [6].

The human defensin family consists of 10 antimicrobial peptides. Human alpha defensins 1–4 (HNP1–4), also called human neutrophil peptides, are mainly secreted from neutrophils [7]. The HNP1–4 possess broad spectrum antibacterial activity against pathogens and modulates innate immunity against infections [2]. Human alpha defensins (HD5 and HD6) are expressed in Paneth cells of the human duodenum, jejunum, and ileum [7]. HD-5 are not expressed in the normal adult colon due to the lack of Paneth cells, but it can be found in metaplastic Paneth cells in the colon of IBD patients [7,8]. *NOD2* genotype is not associated with Paneth cell alpha defensin expression [9]. However, *NOD2* mutations may be associated with reduced ileal expression of HD-5 and HD-6 in Crohn's disease (CD) patients [10]. HD-5 is bactericidal toward *E. coli*, *S. aureus*, and *Salmonella typhimurium* [11–14] and induces IL-8 expression in intestinal epithelial cells.

Constitutive expression of human beta defensin 1 (HBD-1) in both ileal and colonic epithelium is not altered in colitis [2]. Human colonic beta-defensin 2 (HBD-2) expression is significantly elevated in the inflamed colonic epithelium of inflammatory bowel disease (IBD) patients [15]. HBD-2 is very effective in killing *E. coli*, *Pseudomonas aeruginosa*, and *Candida albicans* [16]. HBD-3 and HBD-4 expression are increased in colonic crypts of ulcerative colitis (UC), but not CD patients [17].

Elafin is an antimicrobial peptide with antiprotease activity [18]. It is antibacterial against *P. aeruginosa* and *S. aureus* [19]. Another antiprotease, AMP is a secretory leukocyte protease inhibitor (SLPI), which is expressed in human jejunum and colonic biopsies. SLPI is detrimental to *E. coli* and mycobacteria [20,21].

Lipocalin-2 (LCN2), which is also known as neutrophil gelatinase-associated lipocalin (NGAL), siderocalin, or 24p3 [22], is an antimicrobial peptide belonging to the lipocalin family. LCN2 is expressed in intestinal epithelial cells and myeloid cells [23] and can be found in circulation in both humans and mice [24]. Studies from Goetz et al. showed that LCN2 binds to the iron-bound siderophore, a protein that is involved in iron sequestration, and thus interferes with iron metabolism and growth of *E. coli* [25].

Secreted phospholipase A_2 (sPLA_2) is a peptide that is secreted during digestion and inflammation in the gastrointestinal tract and involved in antimicrobial activity [26]. sPLA_2 is found along the digestive tract [27] and liver [28].

Gastrointestinal AMPs not only protect GI tract from infection and inflammation, but also modulate metabolic activity. In this chapter, we focus on the roles of AMPs found in the gastrointestinal tract, encompassing the esophagus to rectum. We also discuss novel approaches of utilizing gastrointestinal AMPs as biomarkers and drug targets.

2 THE UTILITY OF AMPs AS IBD BIOMARKERS

IBD, that is, UC and CD, are complex autoimmune disorders associated with various symptoms and variable responses to therapy influenced by many factors. Diagnosis and evaluation of disease activity involve ileocolonoscopy, magnetic resonance imaging (MRI), and computed tomography (CT) scans. These procedures are not performed frequently because they may be invasive, risky, and expensive [29,30]. As a result, biomarkers are routinely used to monitor clinical disease activity. C-reactive protein (CRP), erythrocyte sedimentation rate (ESR), and fecal calprotectin are commonly used as IBD biomarkers [31].

CRP is not a specific biomarker of IBD. It is produced in the liver and reflects a general state of inflammation [32]. It has a short half-life (19 h), making it relatively responsive to a rapid change of inflammation. Healthy subjects have less than 1 mg/L of CRP, and acute inflammation can significantly increase this level [33]. CRP is widely used in the diagnosis and monitoring of IBD because it is inexpensive, minimally invasive, and quick to produce results. High CRP levels can indicate active endoscopic disease activity [34] and has been associated with clinical relapse in CD patients [35]. However, some CD patients with active disease show normal levels of CRP; therefore, CRP has a poor correlation with the Harvey Bradshaw Index in CD patients [36]. In UC patients, elevated CRP levels correlate with severe clinical disease activity but not endoscopic disease activity [37]. High CRP levels predict high risks of surgery in both UC and CD patients [38]. ESR is not a specific biomarker of IBD, but it can be used to assess IBD disease activity; however, it is less frequently used than CRP [32]. ESR is not as responsive as CRP in reflecting the acute change in inflammation [39].

Fecal calprotectin (FC) is a reflection of neutrophil inflammatory response in the intestine [40]. FC tests are noninvasive, easy to perform, and resistant to degradation even at room temperature [41]. It is useful to differentiate IBD from Irritable Bowel Syndrome (IBS) [42]. In a systemic review, FC has been shown to be more sensitive than CRP in the assessment of endoscopic activity in the IBD patients [34]. A study demonstrated that FC above 150 g/g of stool indicates increased (14-fold) risk of relapse in UC, but not CD patients [43]. IBD patients with high calprotectin levels after surgery have a high chance of recurrence [44]. However, FC has moderate or little correlation with clinical disease activity in CD patients [45].

Stool lactoferrin (SL) is the first clinically available antimicrobial peptide IBD biomarker [46], which has been shown to correlate closely with mucosal disease activity, especially with mucosal healing [34]. SL shares similar advantages with

FC with respect to convenience and noninvasiveness. Both FC and SL levels have been shown to correlate with endoscopic disease activity [47].

Approximately 10%–25% of CD patients develop at least one intestinal stricture, which are usually diagnosed by an imaging procedure [48,49]. There is no established biomarker available for indicating intestinal stricture in CD. A large-scale single nucleotide polymorphism (SNP) study showed that polymorphism of *NOD2*, *JAK2*, and *ATG16L1* might be associated with intestinal fibrosis [50]. On the other hand, anti-CBir1 and anti-*Saccharomyces cerevisiae* antibodies correlate with stricture development in pediatric and adult patients with CD [51,52]. However, none of these have high specificity in indicating stricturing [53]. Also, no biomarkers have been proposed for indicating fistulas in the literature.

Currently, the only antimicrobial peptides used as clinical IBD biomarkers are FC and SL. Other antimicrobial peptides such as serum hepcidin [54] and beta-defensin-2 [55] do not correlate with IBD disease activity. Schauber et al. demonstrated that colonic cathelicidin (*CAMP*) mRNA expression is increased in ulcerative colitis but not Crohn's disease patients [56]. Dr. Koon's laboratory utilized a cohort of human colonic biopsies from IBD patients out of Cedars-Sinai Medical Center (CSMC) for determining the correlation between cathelicidin expression and IBD disease development.

The human colonic biopsies were obtained after informed consent in accordance with procedures established by the Cedars-Sinai Medical Center's (CSMC) Institutional Review Board, IRBs 3358 and 23,705, and UCLA Institutional Review Board, IRB-11-001527. Inclusion criteria: Involved UC and CD colonic biopsies (IBD groups) or non-IBD colonic biopsies (control groups) from surgical resection from male or female patients, who were able to make informed consent independently, were included. Control group samples were obtained from non-IBD patients with colorectal cancer/polyp, diverticulitis, or colon mass. Exclusion criteria: Pregnant women, prisoners, or minors under age 18 were not included. Baseline characteristics were outlined in Tables 1 and 2. UC clinical disease activity (simple clinical colitis activity) [57], CD clinical disease activity (Harvey Bradshaw Index) [58], and histology score [59] were evaluated according to previously reported approaches.

In the CSMC cohort, the trend of cathelicidin mRNA expression in colons of IBD patients was similar to the report by Schauber et al. (Fig. 1A) [56]. Colonic cathelicidin mRNA expression in UC patients, but not CD patients, was significantly increased. There was no correlation between colonic cathelicidin mRNA expression with histology scores and simple clinical colitis activity scores in UC patients (Fig. 1B and C). In CD patients, there was also no correlation between colonic cathelicidin mRNA expression and histology scores (Fig. 1D). Colonic cathelicidin mRNA expression appeared to be directly proportional to HBI values in CD patients (Fig. 1E). When the cohort was divided into three equal tertiles depending on cathelicidin mRNA expression, CD patients in the high colonic cathelicidin mRNA expression tertile had higher HBI values than CD patients in the lower two tertiles; however, the difference was not statistically significant (Fig. 1F).

TABLE 1 Baseline Characteristics of Cohort 1 and 2, Sorted by Disease Category

	Non-IBD	UC	CD
CAMP mRNA expression (fold)	2.3±0.67	4.3±0.82	3.1±0.48
Colonic LL-37 (ng/μg protein) (mean±SEM)	25±3.5	31±3.8	27±3.9
FPRL1 mRNA expression (fold)	1.7±0.31	3.2±0.65	7.1±2.55
Age at collection (mean±SEM)	60±2.2	41±2.1	40±2.1
Gender (% male)	73	55	73
Histology score (mean±SEM)	2.6±0.3	7.5±0.4	8.4±0.4
COL1A2 mRNA expression (fold)	4.6±0.6	9.3±1.2	6.9±1.3
TGF-b1 mRNA expression (fold)	2.4±0.4	3.1±0.4	1.6±0.2
Vimentin mRNA expression (fold)	1.7±0.2	2.6±0.3	2.1±0.3
n	40	50	44

Data were sorted by non-IBD, UC, and CD.

The CD patients in the middle-to-high tertiles had higher risks of stricture compared to those in the low tertile (Fig. 2A). The details of prevalence, sensitivity, specificity, positive predictive value (PPV), and negative predictive value (NPV) data of using colonic cathelicidin mRNA (*CAMP*) expression for indicating intestinal stricture are shown in Fig. 2B.

Data from the International IBD Genetics Consortium published genome-wide association studies (GWAS) and Immunochip meta-analysis identified a cathelicidin *CAMP* gene single nucleotide polymorphism (SNP) associated with UC (Fig. 3A). UC patients from cohort 1 have high colonic cathelicidin mRNA expression, compared to the control group (Fig. 3B). Colons of UC patients with homozygous TT and heterozygous GT genotype had significantly increased *CAMP* mRNA expression while those with homozygous GG genotype had impaired *CAMP* mRNA expression (Fig. 3B). This finding may partially address the difference of colonic cathelicidin mRNA expression in some UC patients (Fig. 3B). Although the GG genotype of *CAMP* gene contributes to low colonic *CAMP* mRNA expression in UC patients (Fig. 3B), the prevalence of this genotype is rare. It is unlikely to play a significant role in the modulation of colonic disease activity. SNP of *CAMP* gene is not associated with colonic cathelicidin expression in CD patients (data not shown).

Circulating cathelicidin is relevant in autoimmune diseases as well as evidenced by elevated serum cathelicidin levels in psoriasis and anti-neutrophil cytoplasmic antibody-associated vasculitis patients [60,61]. In a recent report, Koon et al. utilized serum samples from two IBD patients for determining the

TABLE 2 Baseline Characteristics of Cohort 1 and 2, Sorted by CAMP mRNA Expression

	Low	Middle	High
Ulcerative colitis			
Range of CAMP mRNA expression (fold)	0–0.8	0.9–4.1	4.4–28.9
Colonic LL-37 (ng/µg protein) (mean±SEM)	25±4.67	35±8.64	33±7.15
Age at collection (mean±SEM)	43±4.3	45±3.42	43±3.1
Gender (% male)	65	44	50
Percentage who used biologics	38	25	7
Percentage who used 6MP or steroids	50	69	33
Duration of disease in years (mean±SEM)	16±4	15±3	13±4
CRP levels (mg/L) (mean±SEM)	3.5±1.14	1.9±0.54	3.8±1.4
Simple clinical colitis activity (mean±SEM)	5.2±1.02	7.2±0.81	6.7±1.21
Histology score (mean±SEM)	7.7±0.66	6.5±0.60	8.6±0.80
n	17	16	17
Crohn's disease			
Range of CAMP mRNA expression (fold)	0–1.0	1.1–2.7	3.1–13.9
Colonic LL-37 (ng/µg protein) (mean±SEM)	27±8.3	26±8.8	30±6.0
Age at collection (mean±SEM)	49±3.6	44±3.6	40±3.5
Gender (% male)	60	71	77
Percentage who used biologics	27	62	8
Percentage who used 6MP or steroids	36	69	54
Duration of disease in years (mean±SEM)	25±3	16±3	10±3
CRP levels (mg/L) (mean±SEM)	1.9±1.19	2.4±1.23	2.88±0.98
HBI (mean±SEM)	5.4±1.2	5.5±0.65	9.7±1.74
Histology score (mean±SEM)	7.9+0.72	8.3±0.89	9.7±0.9
Presence of stricture (%)	20	29	29
n	15	14	15

Data were sorted by CAMP mRNA expression.

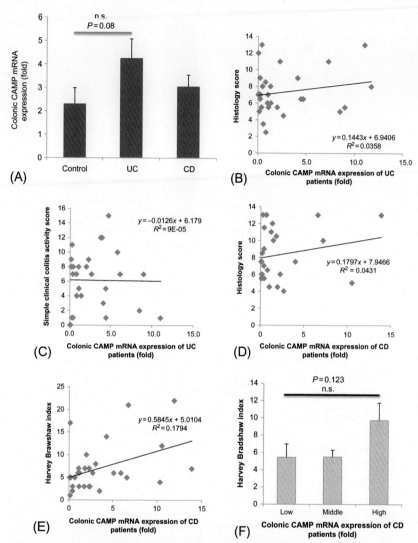

FIG. 1 Colonic *CAMP* mRNA expression is not associated with histology score and clinical disease activity in IBD patients. (A) Colonic *CAMP* mRNA expression of cohort 1. (B) Scatter plot shows the correlation of histology score and colonic *CAMP* mRNA expression in UC patients. (C) Scatter plot shows the correlation of simple clinical colitis activity score and colonic *CAMP* mRNA expression in UC patients. (D) Scatter plot shows the correlation of histology score and colonic *CAMP* mRNA expression in CD patients. (E) Scatter plot shows the correlation of HBI values and colonic *CAMP* mRNA expression of CD patients. (F) The bar graph shows the HBI values of low, middle, and high tertiles of colonic *CAMP* mRNA expression.

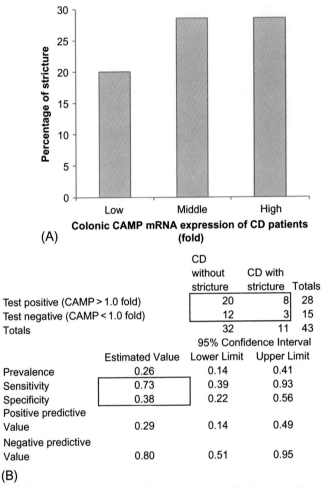

(A)

	CD without stricture	CD with stricture	Totals
Test positive (CAMP > 1.0 fold)	20	8	28
Test negative (CAMP < 1.0 fold)	12	3	15
Totals	32	11	43

		95% Confidence Interval	
	Estimated Value	Lower Limit	Upper Limit
Prevalence	0.26	0.14	0.41
Sensitivity	0.73	0.39	0.93
Specificity	0.38	0.22	0.56
Positive predictive Value	0.29	0.14	0.49
Negative predictive Value	0.80	0.51	0.95

(B)

FIG. 2 Increased colonic *CAMP* mRNA expression is associated with increased occurrence of intestinal stricture in Crohn's disease patients. (A) Percentage of stricture occurrence of low, middle, and high tertiles of colonic *CAMP* mRNA expression. (B) A table shows sensitivity and specificity of using *CAMP* mRNA expression for indicating the presence of stricture.

correlation between serum LL-37 levels and IBD disease activity [62]. Serum cathelicidin levels were inversely proportional to UC clinical disease activity (Partial Mayo Scores) and CD clinical disease activity (Harvey Bradshaw Indices) [62]. The IBD patients with moderate or severe initial clinical disease activity and high initial serum LL-37 levels tend to have significantly better recovery than those with low initial serum LL-37 levels [62]. For UC patients, co-evaluation of LL-37 levels and CRP levels are more accurate in indicating endoscopic disease activity than LL-37 alone or CRP alone. Moreover, low serum LL-37 levels are associated with significantly elevated risk of intestinal

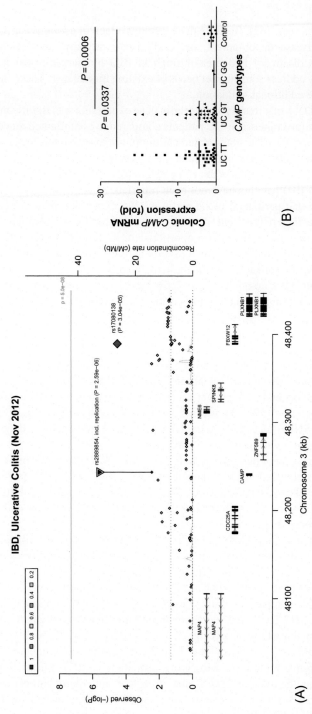

FIG. 3 TT and GT, but not GG, cathelicidin SNP genotype have increased colonic *CAMP* mRNA expression in UC patients. (A) Genome-wide associated study of the *CAMP* gene among UC patients. (B) Colonic *CAMP* mRNA expression in UC patients. Only UC patients with homozygous TT ($P = 0.0337$) and heterozygous GT ($P = 0.0006$) genotypes had significantly increased colonic *CAMP* mRNA expression, compared to control (non-IBD) patients. UC patients with homozygous GG genotype had no altered colonic *CAMP* mRNA expression. The results represented 25 TT homozygous, 15 GT heterozygous, and 2 GG homozygous genotype patients.

stricture in CD patients [62]. Because there remains a lack of a biomarker indicating future disease development, mucosal disease activity, and stricture occurrence, this discovery is a breakthrough in IBD biomarker research as cathelicidin may address some unmet needs of existing biomarkers. These findings will require additional validation.

In an early cDNA macroarray study, elafin gene expression was significantly increased in inflamed mucosa of UC patients compared to noninflamed mucosa of the same patient [63]. The upregulated elafin protein expression is located in enterocytes, goblet cells, and immune cells of the affected mucosa [64]. Mucosal expression of elafin is reduced in the CD patients [65]. Although serum elafin protein levels in IBD patients have not been reported, a small Chinese patient cohort indicated that peripheral blood elafin mRNA expression in active IBD patients are lower than IBD patients in remission [66]. Also, the peripheral blood elafin mRNA expression is inversely proportional to UC clinical disease activity (modified Mayo score) and CD clinical disease activity (CD activity index) in IBD patients [66]. The utility of serum elafin as an IBD biomarker is worth further investigation. Mice do not have the elafin gene, so it is impossible to study how endogenous elafin modulates colitis using mouse models.

Bactericidal/permeability increasing protein (BPI) expression is elevated in the colonic mucosa of UC patients compared to control patients. Tissue levels BPI protein are correlated with disease activity in UC [67]. Some IBD patients have an elevation of anti-neutrophil cytoplasmic (ANCA) IgG antibodies, which can neutralize the antimicrobial effects of BPI [68]. BPI-targeting autoantibodies are associated with greater mucosal damage and severity of symptoms in IBD patients [69]. On the other hand, one *BPI* single nucleotide polymorphism (SNP) (GLU216Lys) has been associated with impaired defense against gram-negative bacteria in CD patients [70].

Colonic expression of human beta-defensin 2 (HBD-2) is very low in a normal colon but is elevated in the inflamed colonic epithelium of IBD patients [15]. Plasma levels of HBD-2 are not altered in IBD patients [71]. HBD-3 and HBD-4 expression are significantly increased in colonic crypts of UC, but not CD patients [17].

Colonic, but not ileal, epithelial cells of UC patients have significantly higher lysozyme mRNA expression than control patients [72]. Fecal lysozyme expression is positively correlated with disease activity [73].

Several studies have identified LCN2 as a potential fecal biomarker [74–76]. LCN2 has a highly compact structure, which makes it resistant to protease degradation [77]. LCN2 levels are increased in colonic mucosal tissues in IBD patients [78]. In one study involving 23 UC patients, 14 CD patients, and 20 healthy volunteers, increased fecal LCN2 expression was found in IBD patients (183 ng/mg protein vs 546 ng/mg protein, $P < 0.01$). Furthermore, when comparing the fecal LCN2 levels of active and quiescent patients, increased fecal LCN2 levels were increased with the severity of UC and CD [76]. Importantly, LCN2 has been directly compared to the well-established fecal biomarker, calprotectin (FC) for its potential to be a fecal biomarker. In a study involving 44 UC and 29 CD and 23 healthy controls, fecal LCN2 levels were also found to be significantly elevated in colonic tissues from patients with active UC and CD. Also, fecal LCN2 levels showed strong correlations with FC ($\rho = 0.82$), endoscopic score and clinical scores and the sensitivity and specificity were 94.7% and 95.7%, respectively [74]. Thus, LCN2 may be a promising fecal IBD biomarker.

3 BACTERIAL-DERIVED SUBSTANCES ACTIVATE INNATE MUCOSAL IMMUNITY VIA CATHELICIDIN EXPRESSION

Cathelicidin is an antimicrobial peptide with reported antiinflammatory effects [1]. Animal studies show that mice with a cathelicidin deficiency are more susceptible to DSS-mediated colitis than wild-type mice [79]. Administration of exogenous cathelicidin not only ameliorates DSS-mediated colitis but also TNBS-intestinal fibrosis [80,81]. These findings support the association of cathelicidin expression with colonic inflammation. Endogenous cathelicidin expression appears to be regulated by bacteria. DSS-exposed mice develop colitis with induction of colonic cathelicidin expression. However, TLR9-deficient mice developed more severe colitis than wild-type mice due to reduced colonic cathelicidin expression [79]. Bone marrow transplantation of wild-type hematopoietic cells ameliorated DSS colitis in TLR9-deficient mice. Bacteria DNA binds to the TLR9 receptor and induces cathelicidin expression [79,82]. The endogenous cathelicidin expression serves as a self-protective system against colitis development.

Intestinal bacteria break down fiber materials to generate short-chain fatty acids. Short-chain fatty acids such as butyrate are the energy sources for colonic epithelial cells [83]. They also have anti-inflammatory effects. Sodium butyrate induces cathelicidin [84] and ameliorates shigellosis [85] but whether increased endogenous cathelicidin is sufficient to inhibit colitis is not known.

To understand the role of butyrate-mediated endogenous cathelicidin induction in the development of acute colitis, mCRAMP deficient ($Camp^{-/-}$) and WT mice were treated with DSS and sodium butyrate as illustrated in Fig. 4A. DSS colitis led to weight loss (Fig. 4B) and significant tissue damage (Fig. 4C) with increased histology score (Fig. 4D). $Camp^{-/-}$ mice had significantly worse colitis than wild-type mice when exposed to DSS (Fig. 4C and D). Interestingly, intraperitoneal sodium butyrate administration led to significantly decreased histology score in wild-type but not $Camp^{-/-}$ deficient mice (Fig. 2D). Also, sodium butyrate administration caused a greater decrease in colonic TNFα levels in WT mice (40% reduction) than $Camp^{-/-}$ mice (21% reduction) (Fig. 5A). Sodium butyrate administration significantly increased colonic Camp mRNA expression in mice (Fig. 5B).

We also treated the WT and $Camp^{-/-}$ mice with TNBS to confirm this finding in another mouse model of colitis (Fig. 6A). Similarly, intraperitoneal

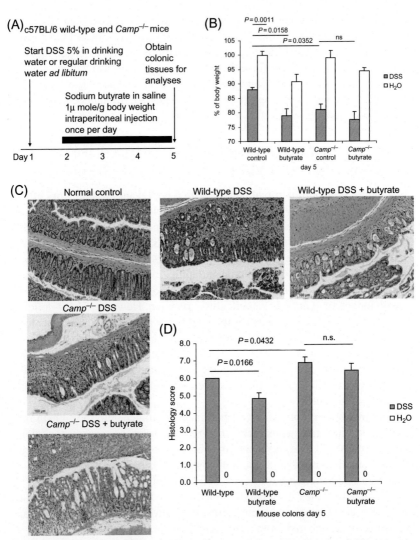

FIG. 4 Sodium butyrate ameliorated DSS mediated colitis via endogenous cathelicidin induction in mice. (A) Illustration of the experimental plan. (B) Body weight change. DSS treatment significantly reduced body weight ($P=0.0011$), compared to regular water treated normal control group. $Camp^{-/-}$ mice developed significantly lower body weight ($P=0.0352$) than wild-type mice after DSS treatment. Sodium butyrate significantly reduced body weight in wild-type mice ($P=0.0158$) but not in $Camp^{-/-}$ mice. (C) H&E staining of the colons. (D) Histology score. DSS treatment led to substantial colonic tissue damages as reflected by increased histology score. $Camp^{-/-}$ mice had worse tissue damages and higher histology score than wild-type mice after exposure to DSS ($P=0.0432$). Sodium butyrate administration significantly reduced histology score of wild-type mice ($P=0.0166$) but not in $Camp^{-/-}$ mice. Experiments included six mice per group.

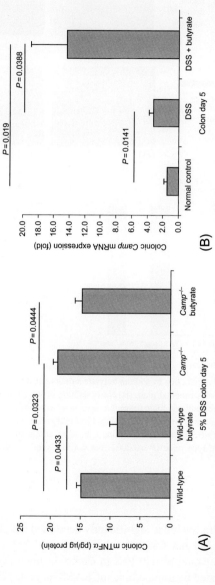

FIG. 5 Sodium butyrate inhibited colonic TNFα expression. (A) Colonic TNFα protein levels. (A) Colonic TNFα protein levels in DSS exposed mice. DSS treated $Camp^{-/-}$ mice had significantly higher colonic TNFα protein expression ($P=0.0323$) than DSS treated WT mice. Colonic TNFα levels in both wild-type and $Camp^{-/-}$ mice were significantly reduced by butyrate treatment ($P=0.04$). The decrease of TNFα levels in $Camp^{-/-}$ mice (drop by 4 pg/μg protein) was smaller than the decrease in wild-type mice (drop by 6 pg/μg protein). (B) Colonic *Camp* mRNA expression. DSS treatment significantly induced colonic *Camp* mRNA expression that was further augmented by sodium butyrate treatment. Experiments included six mice per group.

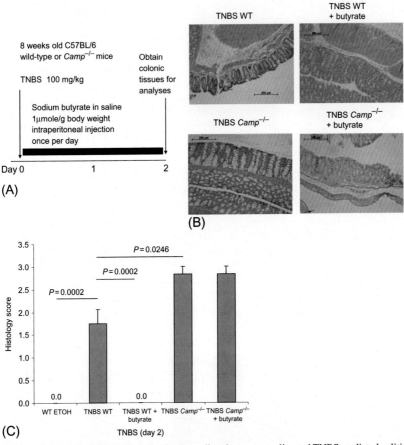

FIG. 6 Endogenous cathelicidin induction by sodium butyrate ameliorated TNBS mediated colitis in mice. (A) Illustration of the experimental plan. (B) H&E staining of the colons. (C) Histology score. TNBS treatment led to substantial colonic tissue damages as reflected by increased histology score ($P=0.0002$). $Camp^{-/-}$ mice had worse tissue damages and higher histology score than wild-type mice after exposure to TNBS ($P=0.0246$). Sodium butyrate administration significantly reduced histology score of wild-type mice ($P=0.0002$) but not in $Camp^{-/-}$ mice. Experiments included six mice per group.

sodium butyrate administration led to significantly decreased histology score in TNBS treated wild-type but not $Camp^{-/-}$ deficient mice (Figs. 6B and C). TNBS colitis resulted in worse weight loss in $Camp^{-/-}$ mice than wild-type mice, which was ameliorated by intracolonic mCRAMP peptide administration (Fig. 7A). Endogenous cathelicidin deficiency or mCRAMP peptide administration did not influence body weight in control ethanol-treated groups (Fig. 7B). Colonic endogenous cathelicidin expression was significantly

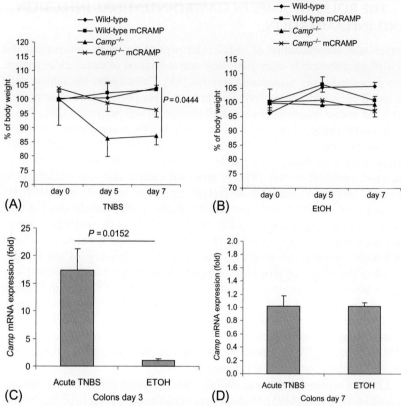

FIG. 7 $Camp^{-/-}$ mice have lower body weight than WT mice in TNBS colitis, and colonic endogenous cathelicidin expression is transient in acute TNBS colitis. (A) Body weight change of TNBS treated groups. $Camp^{-/-}$ mice had significantly lower body weight ($p=0.0444$) than wild-type mice on day 7 after TNBS treatment. (B) Body weight change of ethanol-treated normal groups. mCRAMP treatment did not affect body weight in mice without colitis. (C and D) Colonic $Camp$ mRNA expression of WT mice. Exposure to TNBS significantly increased colonic $Camp$ mRNA expression on day 3 ($P=0.0152$) but not day 7. Experiments included six mice per group.

increased during the early inflammatory stage (day 3) and then restored to baseline level during the late inflammatory stage (day 7) (Figs. 7C and D).

Both studies indicate that butyrate ameliorates acute colitis in mice via induction of endogenous cathelicidin expression. Again, the interactions between butyrate and cathelicidin expression suggest the cooperation of bacterial metabolism and innate immunity. However, there is no evidence suggesting the induction of other antimicrobial peptides by butyrate. Besides bacteria DNA and butyrate, many other bacterial-derived substances such as toll-like receptor agonists can induce mucosal expression of AMPs [86,87].

4   THE ROLES OF AMPs IN GASTROINTESTINAL INFECTION AND INFLAMMATION

Intracolonic administration of cathelicidin peptide has been shown to inhibit *C. difficile* infection in mice [88]. Ileal administration of cathelicidin peptide also prevented toxin A-mediated enteritis [88]. Cathelicidin can inhibit toxin A- and B-mediated TNFα expression in human peripheral blood mononuclear cells and mouse macrophages via inhibition of NF-κB activation [88].

In another study, systemic lentiviral cathelicidin overexpression inhibited Salmonella-mediated cecal inflammation and fibrosis [81]. Cathelicidin-deficient mice are highly susceptible to *E. coli* O157: H7 infection due to an impaired intestinal barrier [89]. A modified cathelicidin (cathelicidin-WA) increased intestinal tight junction protein expression in the *E. coli* O157: H7 infected mice and improved survival [90]. Additionally, cathelicidin can bind to exotoxin lipopolysaccharide (LPS), neutralize its toxicity, and reduce subsequent inflammatory responses [91]. Through these mechanisms, oral delivery of mCRAMP-encoding *Lactococcus,* and intrarectal delivery of mCRAMP-encoding plasmids, can protect mice against DSS-induced colitis [80,92].

N8 Medical, Inc. developed a cathelicidin mimic called Ceragenin Cationic Steroid Antimicrobial 13 (CSA13). It is a nonpeptide-based mimic of cathelicidin with excellent chemical stability and low hemolytic activity [93]. Intravenous or intraperitoneal injection of CSA13 effectively inhibited *P. aeruginosa* (PAO1) infection [94]. It is possible for CSA13 to be a potential candidate drug for treating the gastrointestinal infection.

Elafin, a neutrophil protease inhibitor, has been cloned into food-grade lactic acid bacteria (LAB) [95]. Oral delivery of elafin-expressing LAB inhibits the mouse model of colitis with the protection of an intestinal barrier and inhibition of cytokine and chemokine expression [96]. Another study showed that inactivation of a housekeeping HtrA gene in the elafin-expressing *L. lactis* could enhance elafin expression and improve its protective effect against colitis [97]. Similarly, oral administration of serine protease inhibitor (SLPI) expressing *Lactococcus lactis* are better than IL-10- or TGF-β1-expressing bacteria in treating mouse colitis [97]. While SLPI exerts direct antimicrobial effects against *S. typhimurium*, it does not affect epithelial barrier integrity specifically against *Salmonella* infection [98]. Besides lactic acid bacteria, probiotic *E. coli* Nissle 1917 has recently been developed as a vector for delivering antimicrobial peptides to the site of gastrointestinal infection [99].

In different mouse colitis models, Lcn2 expression was shown to be TLR4-[100,101] and MyD88- [102] dependent. Interestingly, an increase in Lcn2 levels during DSS-induced colitis and colitis induced by neutralization of IL-10 was attenuated by antibiotic treatment, suggesting that the presence of microbiota is essential to Lcn2 expression in mice [102]. LCN2 modulates intestinal inflammation through several mechanisms. In colon tissues infected by *S. typhimurium*, Lcn2 stabilizes MMP-9, a metalloproteinase, and thus exacerbates tissue damage in colonic mucosa [100]. In a Lcn2/Il-10 double-knockout, mice infected by *E. coli*, and mice with Lcn2-repleted macrophages had reduced severity of intestinal inflammation, possibly due to the enhanced phagocytic bacterial clearance in macrophages [101]. On the other hand, Singh et al. showed that while myeloperoxidase (MPO) activity was inhibited by enterobactin (Ent), an *E. coli*-derived siderophore, the binding of Lcn2 to Ent partially restored MPO activity in colon tissues [103]. Along with this line, global Lcn2 knockout mice were shown to have elevated entA-expressing gut bacteria [102]. Thus, Lcn2 modulates intestinal inflammation by regulating macrophage function and directly interfering with bacterial activity in colon tissues.

sPLA$_2$ is produced in Paneth cells and macrophages in small intestine upon proinflammatory stimulation [104]. PLA$_2$ isolated from small intestine was bactericidal against *Listeria monocytogenes*, *E. coli,* and *S. typhimurium* [105]. Also, sPLA$_2$-IID, a member of sPLA$_2$, was highly induced in regulatory T (T_{reg}) cell population and a potential mediator of T_{reg} function. Importantly, administration of recombinant sPLA$_2$-IID protein blocked colitis development in a T-cell transfer colitis mouse model with reduced cytokine production in CD4$^+$ T cells [106].

5   THE ROLES OF AMPs IN OBESITY, LIVER STEATOSIS, AND METABOLIC SYNDROMES

More than 1/3 of American adults and 17% of American children and adolescents are obese [107]. Obesity is also associated with type II diabetes, cardiovascular diseases, nonalcoholic fatty liver disease (NAFLD), and cancer [108]. Hepatic steatosis is affecting 20% American population [109]. Hepatic steatosis, being asymptomatic initially, can progress to steatohepatitis and liver fibrosis if not treated [110]. End-stage fatty liver patients may require liver transplantation [111]. The importance of AMPs in the development of obesity and liver diseases is gaining attention in the scientific community. In some cases, the expression of AMPs in the gastrointestinal tract is relevant to the development of obesity and liver diseases [112].

Koon et al. utilized serum samples from two cohorts of obese patients for determining the correlation between serum LL-37 levels and obesity [113]. In nondiabetic obese patients, serum LL-37 levels are directly proportional to the body mass index (BMI) values. Interestingly, serum LL-37 levels in prediabetic patients are lower than nondiabetic subjects [113]. There is no correlation between serum LL-37 levels and BMI values in type II diabetic patients.

Systemic lentiviral cathelicidin overexpression significantly inhibited hepatic steatosis and reduced fat mass in high-fat diet-treated diabetic mice [113]. This antiobesity effect is mediated by inhibition of fatty acid receptor CD36 expression in adipocytes and hepatocytes. The antiinflammatory effect of cathelicidin in also involved in the protection against diabetic neuropathy. Cathelicidin inhibits aldose reductase and proinflammatory cytokine TNFα expression in peripheral nerves of diabetic mice [113]. Cathelicidin may be a promising drug target in obesity and diabetes.

Insulin-secreting β-cells express cathelicidin [114]. The β-cell cathelicidin expression is regulated by short-chain fatty acids produced by intestinal microbiota. In type 1 diabetic-prone rats, cathelicidin gene expression is reduced in islets [114]. Cathelicidin peptide treatment promoted the abundance of *Actinobacteria* (phylum), *Coriobacteria* (class), *Coriobacteriales* (order), *Coriobacteriaceae* (family), and *Adlercreutzia* (genus), and *Lactobacillus* in the intestine as well as β-cell regeneration in the pancreas [114]. Cathelicidin may help type I diabetic patients restore normal glucose regulation. Gut microbiota can regulate insulin expression via cathelicidin.

As nonalcoholic steatohepatitis (NASH) can lead to hepatic cancer, cathelicidin, as a formyl peptide receptor agonist, switch macrophage polarization from proinflammatory M1 type to M2 type. This M2 type may be associated with exacerbation of hepatocellular carcinoma cells invasion. The impact of this effect on the development of NAFLD and liver cancer still need further investigation [115].

Plasma alpha-defensin is also positively correlated with BMI values in a cohort of 250 children [116]. In a longitudinal study, increased α-defensin levels at age 7 predict the high BMI and waist line at age 10. In the same study, plasma levels of another AMP bacterial/permeability-increasing protein are inversely correlated with BMI values [116]. In NASH patients, the serum neutrophil α-defensin is correlated with hepatic necroinflammation as reflected by NAFLD activity scores (NAS) [117].

Some obese and NAFLD patients tend to have vitamin D deficiency. A recent study demonstrated that vitamin D deficiency is associated with increased pathogenic *Helicobacter hapaticus* and reduced *Akkermansia muciniphila* abundance in the high-fat diet-treated mice [118]. These mice show diminished α-defensin expression in the ileum. Oral administration of α-defensin (DEFA5) reversed gut dysbiosis and ameliorated obesity and hepatic steatosis [118]. Thus, vitamin D can induce ileal α-defensin expression that inhibits obesity and liver disease via modulating intestinal microbiota.

High-fat diet induces SLPI expression in adipose tissues of mice [119]. Bacteria-derived substances are often implicated in the development of obesity. TLR2 and TLR4 ligand peptidoglycan and TLR4 ligand LPS induce SLPI expression in differentiated 3T3-L1 adipocytes. SLPI can inhibit LPS-induced IL-6 expression in the adipocytes. SLPI exerts its antiinflammatory effects via the stabilization of IκBα expression [119].

Circulating lactoferrin levels are low in patients with hyperglycemia and obesity [120]. Neutrophils isolated from type II diabetic patients have reduced LPS-induced lactoferrin expression, but this can be reversed by PPAR agonist rosiglitazone treatment.

Kuefner et al. showed that increased human $sPLA_2$-IIA expression in mice fed a high-fat diet had more efficient blood glucose clearance, less cholesterol in circulation, and higher insulin sensitivity, possibly due to the improved lipid metabolism in the liver [121]. The role of PLA2 in lipid metabolism and development of insulin resistance remained to be elucidated.

6   CONCLUDING REMARKS

AMPs have a variety of functions beyond antimicrobial effects. Most of the AMPs can modulate innate immune responses in epithelial cells, immune cells, and even adipocytes. Some AMPs serve as biomarkers for various diseases. AMPs may also serve as drug targets for new drug development. Recent big data approaches such as microarrays, RNA sequencing, and high-throughput screening platforms, have facilitated the discovery of new AMP functions. Among their many functions, AMPs interact with host cells, but the cellular binding molecules of these AMPs are unknown. It remains unclear whether these AMPs bind to a G-protein-coupled receptor or not. The knowledge of AMP-binding molecules can provide insight into the molecular mechanism of AMPs and development of AMP mimics.

From a therapeutic standpoint, AMPs have toxicity and instability issues. Many natural AMPs cannot be used in humans directly due to short half-life in plasma or hemolytic properties. It is necessary to identify the AMP-binding molecules or receptor and then develop new AMP mimics with improved safety profile and chemical stability.

Nowadays, 16S sequencing is commonly used to more fully understand the gut microbiota [122]. However, most AMP-related gastrointestinal disease studies lack 16S sequencing or metagenomics analysis. As a result, it is difficult to understand how AMPs modulate microbiota in the GI tract.

In gastrointestinal disease states, the gut microbiota and host cells release metabolites that may reflect the altered metabolic pathways. It is necessary to perform untargeted metabolomic analysis of fecal and blood samples to understand how AMPs affect metabolic pathways, which modulate the disease development [123].

The immune activity in diseases often involves complex interactions of immune cells. Most AMP studies only involve a few kinds of innate immune cells such as macrophages, neutrophils, and monocytes. Circulating immune cells should be profiled by flow cytometry-based immunophenotyping [124]. The change of immune cell populations and activation status should be profiled. Many of these cells express cytokines, chemokines, and other mediators. Cell sorting and multiplex ELISA are needed to understand their actions and roles in the gastrointestinal diseases.

Apart from lactoferrin, most AMPs are not orally active. On the other hand, many existing antibiotics and anti-inflammatory agents have enteric formulations. Orally active AMP mimics need to be developed to compete with these existing drugs. If the AMPs do not allow for direct oral administration, special coating or packaging with pH-dependent release may be needed to deliver the AMPs to the sites of infection and inflammation [125]. Pegylation of AMPs may be a good way to protect the AMPs against proteases [126]. Alternatively, food-grade bacteria expressing AMPs may be used [96,127]. However, the safety of these food-grade bacteria need to be further evaluated.

ACKNOWLEDGMENTS

This work was supported by Crohn's and Colitis Foundation of America Career Development Award (#2691) and NIH K01 (DK084256) and R03 (DK103964) grant, and National Center for Advancing Translational Sciences UCLA CTSI Grant (UL1TR001881) to HWK, and United States Public Health Service grant DK046763 to D.Q.S.

Cedar-Sinai Medical Center samples: Clinical data and specimens were provided by the MIRIAD Biobank. MIRIAD is currently supported by F Widjaja Foundation Inflammatory Bowel and Immunobiology Research Institute, NIH grant P01DK046763, The European Union Grant 305479, NIDDK Grant DK062413, U54 DE023798, and the Leona M and Harry B Leona M. and Harry B. Helmsley Charitable Trust.

We thank Dr. Phillip Fleshner, Dr. Lori Robbins, Dr. Michelle Vu, Dr. Tressia C. Hing for providing clinical data and colonic biopsy samples of CSMC cohort.

We thank Prof. Richard Gallo for providing *Camp* deficient breeders to Dr. Koon.

DISCLOSURE

All authors have nothing to disclose. No conflicts of interest exist.

REFERENCES

[1] Ho S, Pothoulakis C, Koon HW. Antimicrobial peptides and colitis. Curr Pharm Des 2013;19:40–7.
[2] Wehkamp J, Schauber J, Stange EF. Defensins and cathelicidins in gastrointestinal infections. Curr Opin Gastroenterol 2007;23:32–8.
[3] Murakami M, Dorschner RA, Stern LJ, et al. Expression and secretion of cathelicidin antimicrobial peptides in murine mammary glands and human milk. Pediatr Res 2005;57:10–5.
[4] Zhang L, Wu WK, Gallo RL, et al. Critical role of antimicrobial peptide cathelicidin for controlling Helicobacter pylori survival and infection. J Immunol 2016;196:1799–809.
[5] Lee CC, Sun Y, Qian S, et al. Transmembrane pores formed by human antimicrobial peptide LL-37. Biophys J 2011;100:1688–96.
[6] Sochacki KA, Barns KJ, Bucki R, et al. Real-time attack on single Escherichia coli cells by the human antimicrobial peptide LL-37. Proc Natl Acad Sci U S A 2011;108:E77–81.
[7] Cunliffe RN. Alpha-defensins in the gastrointestinal tract. Mol Immunol 2003;40:463–7.
[8] Cunliffe RN, Rose FR, Keyte J, et al. Human defensin 5 is stored in precursor form in normal Paneth cells and is expressed by some villous epithelial cells and by metaplastic Paneth cells in the colon in inflammatory bowel disease. Gut 2001;48:176–85.
[9] Simms LA, Doecke JD, Walsh MD, et al. Reduced alpha-defensin expression is associated with inflammation and not NOD2 mutation status in ileal Crohn's disease. Gut 2008;57:903–10.

[10] Wehkamp J, Harder J, Weichenthal M, et al. NOD2 (CARD15) mutations in Crohn's disease are associated with diminished mucosal alpha-defensin expression. Gut 2004;53:1658–64.

[11] Ouellette AJ. Paneth cell alpha-defensins: peptide mediators of innate immunity in the small intestine. Springer Semin Immunopathol 2005;27:133–46.

[12] Ouellette AJ. Paneth cell alpha-defensin synthesis and function. Curr Top Microbiol Immunol 2006;306:1–25.

[13] Ouellette AJ. Paneth cell alpha-defensins in enteric innate immunity. Cell Mol Life Sci 2011;68:2215–29.

[14] Ouellette AJ, Satchell DP, Hsieh MM, et al. Characterization of luminal paneth cell alpha-defensins in mouse small intestine. Attenuated antimicrobial activities of peptides with truncated amino termini. J Biol Chem 2000;275:33969–73.

[15] Wehkamp J, Fellermann K, Herrlinger KR, et al. Human beta-defensin 2 but not beta-defensin 1 is expressed preferentially in colonic mucosa of inflammatory bowel disease. Eur J Gastroenterol Hepatol 2002;14:745–52.

[16] Harder J, Bartels J, Christophers E, et al. A peptide antibiotic from human skin. Nature 1997;387:861.

[17] Fahlgren A, Hammarstrom S, Danielsson A, et al. Beta-defensin-3 and -4 in intestinal epithelial cells display increased mRNA expression in ulcerative colitis. Clin Exp Immunol 2004;137:379–85.

[18] Williams SE, Brown TI, Roghanian A, et al. SLPI and elafin: one glove, many fingers. Clin Sci (Lond) 2006;110:21–35.

[19] Wiesner J, Vilcinskas A. Antimicrobial peptides: the ancient arm of the human immune system. Virulence 2010;1:440–64.

[20] Nishimura J, Saiga H, Sato S, et al. Potent antimycobacterial activity of mouse secretory leukocyte protease inhibitor. J Immunol 2008;180:4032–9.

[21] Miller KW, Evans RJ, Eisenberg SP, et al. Secretory leukocyte protease inhibitor binding to mRNA and DNA as a possible cause of toxicity to Escherichia coli. J Bacteriol 1989;171:2166–72.

[22] Kjeldsen L, Cowland JB, Borregaard N. Human neutrophil gelatinase-associated lipocalin and homologous proteins in rat and mouse. Biochim Biophys Acta 2000;1482:272–83.

[23] Devireddy LR, Gazin C, Zhu X, et al. A cell-surface receptor for lipocalin 24p3 selectively mediates apoptosis and iron uptake. Cell 2005;123:1293–305.

[24] Wang Y, Lam KS, Kraegen EW, et al. Lipocalin-2 is an inflammatory marker closely associated with obesity, insulin resistance, and hyperglycemia in humans. Clin Chem 2007;53:34–41.

[25] Goetz DH, Holmes MA, Borregaard N, et al. The neutrophil lipocalin NGAL is a bacteriostatic agent that interferes with siderophore-mediated iron acquisition. Mol Cell 2002;10:1033–43.

[26] Nevalainen TJ, Graham GG, Scott KF. Antibacterial actions of secreted phospholipases A2. Review. Biochim Biophys Acta 2008;1781:1–9.

[27] Eerola LI, Surrel F, Nevalainen TJ, et al. Analysis of expression of secreted phospholipases A2 in mouse tissues at protein and mRNA levels. Biochim Biophys Acta 2006;1761:745–56.

[28] Kramer RM, Hession C, Johansen B. Et al. structure and properties of a human non-pancreatic phospholipase A2. J Biol Chem 1989;264:5768–75.

[29] Jensen MD, Ormstrup T, Vagn-Hansen C, et al. Interobserver and intermodality agreement for detection of small bowel Crohn's disease with MR enterography and CT enterography. Inflamm Bowel Dis 2011;17:1081–8.

[30] Peloquin JM, Pardi DS, Sandborn WJ, et al. Diagnostic ionizing radiation exposure in a population-based cohort of patients with inflammatory bowel disease. Am J Gastroenterol 2008;103:2015–22.

[31] D'Haens G, Ferrante M, Vermeire S, et al. Fecal calprotectin is a surrogate marker for endoscopic lesions in inflammatory bowel disease. Inflamm Bowel Dis 2012;18:2218–24.

[32] Vermeire S, Van Assche G, Rutgeerts P. Laboratory markers in IBD: useful, magic, or unnecessary toys? Gut 2006;55:426–31.

[33] Fengming Y, Jianbing W. Biomarkers of inflammatory bowel disease. Dis Markers 2014;2014.

[34] Mosli MH, Zou G, Garg SK, et al. C-reactive protein, fecal calprotectin, and stool lactoferrin for detection of endoscopic activity in symptomatic inflammatory bowel disease patients: a systematic review and meta-analysis. Am J Gastroenterol 2015;110:802–19. quiz 820.

[35] Koelewijn CL, Schwartz MP, Samsom M, et al. C-reactive protein levels during a relapse of Crohn's disease are associated with the clinical course of the disease. World J Gastroenterol 2008;14:85–9.

[36] Rodgers AD, Cummins AG. CRP correlates with clinical score in ulcerative colitis but not in Crohn's disease. Dig Dis Sci 2007;52:2063–8.

[37] Solem CA, Loftus Jr EV, Tremaine WJ, et al. Correlation of C-reactive protein with clinical, endoscopic, histologic, and radiographic activity in inflammatory bowel disease. Inflamm Bowel Dis 2005;11:707–12.

[38] Henriksen M, Jahnsen J, Lygren I, et al. C-reactive protein: a predictive factor and marker of inflammation in inflammatory bowel disease. Results from a prospective population-based study. Gut 2008;57:1518–23.

[39] Fagan EA, Dyck RF, Maton PN, et al. Serum levels of C-reactive protein in Crohn's disease and ulcerative colitis. Eur J Clin Invest 1982;12:351–9.

[40] Fagerhol MK, Dale I, Andersson T. A radioimmunoassay for a granulocyte protein as a marker in studies on the turnover of such cells. Bull Eur Physiopathol Respir 1980;16 (Suppl):273–82.

[41] Roseth AG, Fagerhol MK, Aadland E, et al. Assessment of the neutrophil dominating protein calprotectin in feces. A methodologic study. Scand J Gastroenterol 1992;27:793–8.

[42] Menees SB, Powell C, Kurlander J, et al. A meta-analysis of the utility of C-reactive protein, erythrocyte sedimentation rate, fecal calprotectin, and fecal lactoferrin to exclude inflammatory bowel disease in adults with IBS. Am J Gastroenterol 2015;110:444–54.

[43] Costa F, Mumolo MG, Ceccarelli L, et al. Calprotectin is a stronger predictive marker of relapse in ulcerative colitis than in Crohn's disease. Gut 2005;54:364–8.

[44] Lobaton T, Lopez-Garcia A, Rodriguez-Moranta F, et al. A new rapid test for fecal calprotectin predicts endoscopic remission and postoperative recurrence in Crohn's disease. J Crohns Colitis 2013;7:e641–51.

[45] Jones J, Loftus Jr EV, Panaccione R, et al. Relationships between disease activity and serum and fecal biomarkers in patients with Crohn's disease. Clin Gastroenterol Hepatol 2008;6:1218–24.

[46] Langhorst J, Elsenbruch S, Koelzer J, et al. Noninvasive markers in the assessment of intestinal inflammation in inflammatory bowel diseases: performance of fecal lactoferrin, calprotectin, and PMN-elastase, CRP, and clinical indices. Am J Gastroenterol 2008;103:162–9.

[47] Puolanne AM, Kolho KL, Alfthan H, et al. Rapid faecal tests for detecting disease activity in colonic IBD. Eur J Clin Invest 2016;46(10):825–32. https://doi.org/10.1111/eci.12660.

[48] Chang CW, Wong JM, Tung CC, et al. Intestinal stricture in Crohn's disease. Intest Res 2015;13:19–26.

[49] Chan G, Fefferman DS, Farrell RJ. Endoscopic assessment of inflammatory bowel disease: colonoscopy/esophagogastroduodenoscopy. Gastroenterol Clin North Am 2012;41:271–90.

[50] Cleynen I, Gonzalez JR, Figueroa C, et al. Genetic factors conferring an increased suscepti-bility to develop Crohn's disease also influence disease phenotype: results from the IBDchip European project. Gut 2013;62:1556–65.

[51] Targan SR, Landers CJ, Yang H, et al. Antibodies to CBir1 flagellin define a unique response that is associated independently with complicated Crohn's disease. Gastroenterology 2005;128:2020–8.

[52] Amre DK, Lu SE, Costea F, et al. Utility of serological markers in predicting the early occur-rence of complications and surgery in pediatric Crohn's disease patients. Am J Gastroenterol 2006;101:645–52.

[53] Bettenworth D, Nowacki TM, Cordes F, et al. Assessment of stricturing Crohn's disease: cur-rent clinical practice and future avenues. World J Gastroenterol 2016;22:1008–16.

[54] Pakoz ZB, Cekic C, Arabul M, et al. An evaluation of the correlation between Hepcidin serum levels and disease activity in inflammatory bowel disease. Gastroenterol Res Pract 2015;2015.

[55] Kolho KL, Sipponen T, Valtonen E, et al. Fecal calprotectin, MMP-9, and human beta-defensin-2 levels in pediatric inflammatory bowel disease. Int J Colorectal Dis 2014;29:43–50.

[56] Schauber J, Rieger D, Weiler F, et al. Heterogeneous expression of human cathelicidin hCAP18/LL-37 in inflammatory bowel diseases. Eur J Gastroenterol Hepatol 2006;18:615–21.

[57] Walmsley RS, Ayres RC, Pounder RE, et al. A simple clinical colitis activity index. Gut 1998;43:29–32.

[58] Harvey RF, Bradshaw JM. A simple index of Crohn's-disease activity. Lancet 1980;1:514.

[59] D'Haens GR, Geboes K, Peeters M, et al. Early lesions of recurrent Crohn's disease caused by infusion of intestinal contents in excluded ileum. Gastroenterology 1998;114:262–7.

[60] Al-Mutairi N, El Eassa B, Nair V. Measurement of vitamin D and cathelicidin (LL-37) levels in patients of psoriasis with co-morbidities. Indian J Dermatol Venereol Leprol 2013;79:492–6.

[61] Zhang Y, Shi W, Tang S, et al. The influence of cathelicidin LL37 in human anti-neutrophils cytoplasmic antibody (ANCA)-associated vasculitis. Arthritis Res Ther 2013;15:R161.

[62] Tran DH, Wang J, Ha C, et al. Circulating cathelicidin levels correlate with mucosal disease activity in ulcerative colitis, risk of intestinal stricture in Crohn's disease, and clinical prog-nosis in inflammatory bowel disease. BMC Gastroenterol 2017;17:63.

[63] Okahara S, Arimura Y, Yabana T, et al. Inflammatory gene signature in ulcerative colitis with cDNA macroarray analysis. Aliment Pharmacol Ther 2005;21:1091–7.

[64] Flach CF, Eriksson A, Jennische E, et al. Detection of elafin as a candidate biomarker for ulcerative colitis by whole-genome microarray screening. Inflamm Bowel Dis 2006;12:837–42.

[65] Schmid M, Fellermann K, Fritz P, et al. Attenuated induction of epithelial and leukocyte ser-ine antiproteases elafin and secretory leukocyte protease inhibitor in Crohn's disease. J Leukoc Biol 2007;81:907–15.

[66] Zhang W, Teng GG, Tian Y, et al. Expression of elafin in peripheral blood in inflammatory bowel disease patients and its clinical significance. Zhonghua Yi Xue Za Zhi 2016;96:1120–3.

[67] Haapamaki MM, Haggblom JO, Gronroos JM, et al. Bactericidal/permeability-increasing protein in colonic mucosa in ulcerative colitis. Hepatogastroenterology 1999;46:2273–7.

[68] Stoffel MP, Csernok E, Herzberg C, et al. Anti-neutrophil cytoplasmic antibodies (ANCA) directed against bactericidal/permeability increasing protein (BPI): a new seromarker for inflammatory bowel disease and associated disorders. Clin Exp Immunol 1996;104:54–9.

[69] Schinke S, Fellermann K, Herlyn K, et al. Autoantibodies against the bactericidal/permeability-increasing protein from inflammatory bowel disease patients can impair the antibiotic activity of bactericidal/permeability-increasing protein. Inflamm Bowel Dis 2004;10:763–70.

[70] Klein W, Tromm A, Folwaczny C, et al. A polymorphism of the bactericidal/permeability increasing protein (BPI) gene is associated with Crohn's disease. J Clin Gastroenterol 2005;39:282–3.

[71] Yamaguchi N, Isomoto H, Mukae H, et al. Concentrations of alpha- and beta-defensins in plasma of patients with inflammatory bowel disease. Inflamm Res 2009;58:192–7.

[72] Fahlgren A, Hammarstrom S, Danielsson A, et al. Increased expression of antimicrobial peptides and lysozyme in colonic epithelial cells of patients with ulcerative colitis. Clin Exp Immunol 2003;131:90–101.

[73] Hidaka M, Sudoh I, Miyaoka M, et al. Measurement of fecal proteins in inflammatory bowel disease—usefulness as an activity index. Nihon Shokakibyo Gakkai Zasshi 2000;97:161–9.

[74] Thorsvik S, Damas JK, Granlund AV, et al. Fecal neutrophil gelatinase-associated lipocalin as a biomarker for inflammatory bowel disease. J Gastroenterol Hepatol 2017;32:128–35.

[75] Peterson CG, Eklund E, Taha Y, et al. A new method for the quantification of neutrophil and eosinophil cationic proteins in feces: establishment of normal levels and clinical application in patients with inflammatory bowel disease. Am J Gastroenterol 2002;97:1755–62.

[76] Nielsen OH, Gionchetti P, Ainsworth M, et al. Rectal dialysate and fecal concentrations of neutrophil gelatinase-associated lipocalin, interleukin-8, and tumor necrosis factor-alpha in ulcerative colitis. Am J Gastroenterol 1999;94:2923–8.

[77] Peitsch MC, Boguski MS. The first lipocalin with enzymatic activity. Trends Biochem Sci 1991;16:363.

[78] Nielsen BS, Borregaard N, Bundgaard JR, et al. Induction of NGAL synthesis in epithelial cells of human colorectal neoplasia and inflammatory bowel diseases. Gut 1996;38:414–20.

[79] Koon HW, Shih DQ, Chen J, et al. Cathelicidin signaling via the Toll-like receptor protects against colitis in mice. Gastroenterology 2011;141(1852–63):e1–3.

[80] Tai EK, Wu WK, Wong HP, et al. A new role for cathelicidin in ulcerative colitis in mice. Exp Biol Med (Maywood) 2007;232:799–808.

[81] Yoo JH, Ho S, Tran DH, et al. Anti-fibrogenic effects of the anti-microbial peptide cathelicidin in murine colitis-associated fibrosis. Cell Mol Gastroenterol Hepatol 2015;1(55-74):e1.

[82] Brandenburg LO, Jansen S, Albrecht LJ, et al. CpG oligodeoxynucleotides induce the expression of the antimicrobial peptide cathelicidin in glial cells. J Neuroimmunol 2013;255:18–31.

[83] Kelly CJ, Zheng L, Campbell EL, et al. Crosstalk between microbiota-derived short-chain fatty acids and intestinal epithelial HIF augments tissue barrier function. Cell Host Microbe 2015;17:662–71.

[84] Schauber J, Iffland K, Frisch S, et al. Histone-deacetylase inhibitors induce the cathelicidin LL-37 in gastrointestinal cells. Mol Immunol 2004;41:847–54.

[85] Raqib R, Sarker P, Mily A, et al. Efficacy of sodium butyrate adjunct therapy in shigellosis: a randomized, double-blind, placebocontrolled clinical trial. BMC Infect Dis 2012;12:111.

[86] Kim SK, Park S, Lee ES. Toll-like receptors and antimicrobial peptides expressions of psoriasis: correlation with serum vitamin D level. J Korean Med Sci 2010;25:1506–12.

[87] Mu C, Yang Y, Zhu W. Crosstalk between the immune receptors and gut microbiota. Curr Protein Pept Sci 2015;16:622–31.

[88] Hing TC, Ho S, Shih DQ, et al. The antimicrobial peptide cathelicidin modulates Clostridium difficile-associated colitis and toxin A-mediated enteritis in mice. Gut 2013;62:1295–305.

[89] Chromek M, Arvidsson I, Karpman D. The antimicrobial peptide cathelicidin protects mice from Escherichia coli O157:H7-mediated disease. PLoS One 2012;7.

[90] Yi H, Hu W, Chen S, et al. Cathelicidin-WA improves intestinal epithelial barrier function and enhances host defense against Enterohemorrhagic Escherichia coli O157:H7 infection. J Immunol 2017;198:1696–705.

[91] Golec M, Cathelicidin LL. 37: LPS-neutralizing, pleiotropic peptide. Ann Agric Environ Med 2007;14:1–4.

[92] Tai EK, Wu WK, Wang XJ, et al. Intrarectal administration of mCRAMP-encoding plasmid reverses exacerbated colitis in Cnlp(−/−) mice. Gene Ther 2013;20(2):187–93. https://doi.org/10.1038/gt.2012.22.

[93] Moscoso M, Esteban-Torres M, Menendez M, et al. In vitro bactericidal and bacteriolytic activity of ceragenin CSA-13 against planktonic cultures and biofilms of Streptococcus pneumoniae and other pathogenic streptococci. PLoS One 2014;9.

[94] Bucki R, Niemirowicz K, Wnorowska U, et al. Bactericidal activity of Ceragenin CSA-13 in cell culture and in an animal model of peritoneal infection. Antimicrob Agents Chemother 2015;59:6274–82.

[95] Greenhill C. IBD: elafin—a potential IBD therapy. Nat Rev Gastroenterol Hepatol 2012;9:686.

[96] Motta JP, Bermudez-Humaran LG, Deraison C, et al. Food-grade bacteria expressing elafin protect against inflammation and restore colon homeostasis. Sci Transl Med 2012;4.

[97] Bermudez-Humaran LG, Motta JP, Aubry C, et al. Serine protease inhibitors protect better than IL-10 and TGF-beta anti-inflammatory cytokines against mouse colitis when delivered by recombinant lactococci. Microb Cell Fact 2015;14:26.

[98] Si-Tahar M, Merlin D, Sitaraman S, et al. Constitutive and regulated secretion of secretory leukocyte proteinase inhibitor by human intestinal epithelial cells. Gastroenterology 2000;118:1061–71.

[99] Geldart K, Forkus B, McChesney E, et al. pMPES: a modular peptide expression system for the delivery of antimicrobial peptides to the site of gastrointestinal infections using probiotics. Pharmaceuticals (Basel) 2016;9(4):E60.

[100] Kundu P, Ling TW, Korecka A, et al. Absence of intestinal PPARgamma aggravates acute infectious colitis in mice through a lipocalin-2-dependent pathway. PLoS Pathog 2014;10.

[101] Toyonaga T, Matsuura M, Mori K, et al. Lipocalin 2 prevents intestinal inflammation by enhancing phagocytic bacterial clearance in macrophages. Sci Rep 2016;6.

[102] Singh V, Yeoh BS, Chassaing B, et al. Microbiota-inducible innate immune, siderophore binding protein lipocalin 2 is critical for intestinal homeostasis. Cell Mol Gastroenterol Hepatol 2016;2(482–498):e6.

[103] Singh V, Yeoh BS, Xiao X, et al. Interplay between enterobactin, myeloperoxidase and lipocalin 2 regulates E. coli survival in the inflamed gut. Nat Commun 2015;6:7113.

[104] Murakami M, Taketomi Y, Miki Y, et al. Recent progress in phospholipase A(2) research: from cells to animals to humans. Prog Lipid Res 2011;50:152–92.

[105] Harwig SS, Tan L, Qu XD, et al. Bactericidal properties of murine intestinal phospholipase A2. J Clin Invest 1995;95:603–10.

[106] von Allmen CE, Schmitz N, Bauer M, et al. Secretory phospholipase A2-IID is an effector molecule of CD4+CD25+ regulatory T cells. Proc Natl Acad Sci U S A 2009;106:11673–8.

[107] Ogden CL, Carroll MD, Kit BK, et al. Prevalence of childhood and adult obesity in the United States, 2011-2012. JAMA 2014;311:806–14.

[108] Finkelstein EA, Trogdon JG, Cohen JW, et al. Annual medical spending attributable to obesity: payer-and service-specific estimates. Health Aff (Millwood) 2009;28:w822–31.

[109] Fazel Y, Koenig AB, Sayiner M, et al. Epidemiology and natural history of non-alcoholic fatty liver disease. Metabolism 2016;65:1017–25.

[110] Yoon HJ, Cha BS. Pathogenesis and therapeutic approaches for non-alcoholic fatty liver disease. World J Hepatol 2014;6(11):800.

[111] Zezos P, Renner EL. Liver transplantation and non-alcoholic fatty liver disease. World J Gastroenterol 2014;20:15532–8.

[112] Araujo JR, Tomas J, Brenner C, et al. Impact of high-fat diet on the intestinal microbiota and small intestinal physiology before and after the onset of obesity. Biochimie 2017;141:97–106. https://doi.org/10.1016/j.biochi.2017.05.019.

[113] Hoang-Yen Tran D, Hoang-Ngoc Tran D, Mattai SA, et al. Cathelicidin suppresses lipid accumulation and hepatic steatosis by inhibition of the CD36 receptor. Int J Obes (Lond) 2016;40(9):1424–34. https://doi.org/10.1038/ijo.2016.90.

[114] Pound LD, Patrick C, Eberhard CE, et al. Cathelicidin antimicrobial peptide: a novel regulator of islet function, islet regeneration, and selected gut bacteria. Diabetes 2015;64:4135–47.

[115] Li Y, Cai L, Wang H, et al. Pleiotropic regulation of macrophage polarization and tumorigenesis by formyl peptide receptor-2. Oncogene 2011;30:3887–99.

[116] Prats-Puig A, Gispert-Sauch M, Carreras-Badosa G, et al. Alpha-defensins and bacterial/permeability-increasing protein as new markers of childhood obesity. Pediatr Obes 2017;12:e10–3.

[117] Manco M, Fernandez-Real JM, Vecchio FM, et al. The decrease of serum levels of human neutrophil alpha-defensins parallels with the surgery-induced amelioration of NASH in obesity. Obes Surg 2010;20:1682–9.

[118] Su D, Nie Y, Zhu A, et al. Vitamin D signaling through induction of Paneth cell defensins maintains gut microbiota and improves metabolic disorders and hepatic steatosis in animal models. Front Physiol 2016;7:498.

[119] Adapala VJ, Buhman KK, Ajuwon KM. Novel anti-inflammatory role of SLPI in adipose tissue and its regulation by high fat diet. J Inflamm (Lond) 2011;8:5.

[120] Moreno-Navarrete JM, Ortega FJ, Bassols J, et al. Decreased circulating lactoferrin in insulin resistance and altered glucose tolerance as a possible marker of neutrophil dysfunction in type 2 diabetes. J Clin Endocrinol Metab 2009;94:4036–44.

[121] Kuefner MS, Pham K, Redd JR, et al. Secretory phospholipase A2 group IIA (PLA2G2A) modulates insulin sensitivity and metabolism. J Lipid Res 2017;58(9):1822–33. https://doi.org/10.1194/jlr.M076141.

[122] Chen J, Huang C, Wang J, et al. Dysbiosis of intestinal microbiota and decrease in paneth cell antimicrobial peptide level during acute necrotizing pancreatitis in rats. PLoS One 2017;12.

[123] Cui DN, Wang X, Chen JQ, et al. Quantitative evaluation of the compatibility effects of Huangqin decoction on the treatment of irinotecan-induced gastrointestinal toxicity using untargeted metabolomics. Front Pharmacol 2017;8:211.

[124] Renaudineau Y. Immunophenotyping as a new tool for classification and monitoring of systemic autoimmune diseases. Clin Rev Allergy Immunol 2017;53(2):177–80. https://doi.org/10.1007/s12016-017-8604-9.

[125] Zhu Q, Talton J, Zhang G, et al. Large intestine-targeted, nanoparticle-releasing oral vaccine to control genitorectal viral infection. Nat Med 2012;18:1291–6.

[126] Kelly GJ, Kia AF, Hassan F, et al. Polymeric prodrug combination to exploit the therapeutic potential of antimicrobial peptides against cancer cells. Org Biomol Chem 2016;14:9278–86.

[127] Zhang L, Yu J, Wong CC, et al. Cathelicidin protects against Helicobacter pylori colonization and the associated gastritis in mice. Gene Ther 2013;20:751–60.

Chapter 4

Cathelicidin in Gastrointestinal Disorders

Jing Shen and Zhangang Xiao
Southwest Medical University, Luzhou, China

Chapter Outline

1 INTRODUCTION

Cathelicidin is a family of evolutionarily conserved antimicrobial peptides described in mammals, birds, fish, and reptiles. Because of the high similarity of their pro-regions to cathelin, a cathepsin L inhibitor from porcine leukocytes [1], this family was named cathelicidin. Its role as an effector molecule of the innate immunity is widening from solely endogenous antibiotics to multiple functional mediators providing the first line of host defense, modifying the local inflammatory response, and activating adaptive immunity.

Antimicrobial Peptides in Gastrointestinal Diseases. https://doi.org/10.1016/B978-0-12-814319-3.00004-0
Copyright © 2018 Chi Hin Cho. Published by Elsevier Ltd. All rights reserved.

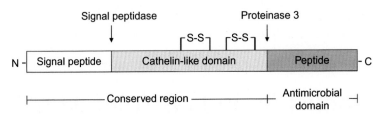

FIG. 1 Schematic representation of the cathelicidin family.

The peptide is synthesized in precursor form with a structurally varied C-terminal domain (Fig. 1). After removal of the signal peptide, it is stored in granules as inactive proform. The active biologic region of cathelicidin resides in the C-terminal. The C-terminal antibacterial peptide is activated upon cleavage by proteases from azurophil granules [2]. Myeloid bone marrow cells are the primary site of expression and it is stored principally in the secretory granules of neutrophils, but a wider distribution has been reported in other cells. For instance, the human cathelicidin, LL-37, is also found in epithelia [3], lymphocytes and monocytes [4]. The mature peptide is released from the storage form in activated neutrophils during degranulation into the phagocytic vacuole or extracellular milieu [5,6]. Because cathelicidin naturally exists and is relatively non-toxic to normal mammalian system, it may have significant clinical implications as therapeutic agents. This chapter will focus on the therapeutic potential of cathelicidin in the gastrointestinal (GI) tract disorders, including inflammation, cancer, and wound repair, with particular emphasis on disorders in the stomach and colon.

2 THE DISTRIBUTION AND FUNCTIONS OF CATHELICIDIN IN THE GI TRACT

2.1 The Distribution of Cathelicidin in the GI Tract

This host defense peptide cathelicidin is predominantly expressed in epithelia, lymphocytes and monocytes [3,4]. The human cathelicidin LL-37 is expressed along the GI tract. In the normal stomach, it is actively produced by surface epithelial cells, as well as chief and parietal cells and is also present in the gastric secretion [7]. In the colon, LL-37is detectable in epithelial cells located at the surface and upper crypts [8]. However, the expression of LL-37 is dysregulated in disease states. In the course of *Helicobacter pylori* (*H. pylori*) infection, LL-37 is induced throughout the whole gastric tubules. During the progression from atrophic gastritis to adenocarcinoma, the expression of LL-37 is reduced [7]. It is absent or expressed at a very low level in gastric hyperplastic polyps, tubular adenomas and adenocarcinomas, suggesting that cathelicidin could play an important role in preventing bacteria-related inflammation and perhaps also carcinogenesis in the GI tract.

2.2 The Functions of Cathelicidin in the GI Tract

The most prominent function of cathelicidin is its ability to inhibit propagation of a diverse range of microorganisms. Experimental evidence also shows that cathelicidin can modulate inflammation and orchestrate wound healing in response to mucosal damage in the GI tract.

2.2.1 Mechanism of Antimicrobial Action

Cathelicidin is microbiocidal against Gram-positive and/or Gram-negative bacteria [9,10], parasites [10,11], fungi [12,13], and enveloped viruses [14] at concentrations in the micromolar range. Some antibiotic-resistant isolates including methicillin-resistant *Staphylococcus aureus* (MRSA), vancomycin-resistant *Enterococcus faecalis* (VREF), and *Pseudomonas aeruginosa* strains from cystic fibrosis patients are also susceptible to the action of cathelicidin [9,15].

The microbiocidal action is primarily determined by their cationic and amphipathic nature. Cathelicidinis highly cationic and have a structure which arranges charged residues separated from hydrophobic one. This structural characteristic enables the peptide to interact electrostatically with the anionic components of bacterial, fungal, vial, and protozoan lipid membranes [15,16]. Following the insertion in the membrane, it may exert bacteriocidal activity by two main mechanisms (Fig. 2): (1) transmembrane pore formation, and (2) membrane destruction, leading to leakage of intracellular components and cell death [17]. In addition to this fundamental killing action, Boman et al. have shown that porcine cathelicidin, PR-39, can suppress microbial growth by inhibiting protein as well as DNA synthesis and finally result in degradation of these components [18].

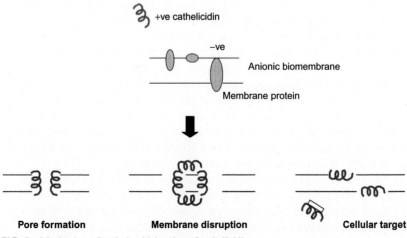

FIG. 2 Mechanism of antimicrobial action of cathelicidin.

2.2.2   Preventing Lipopolysaccharide (Lps)-Induced Inflammation

LPS, a major constituent of the outer membrane of Gram-negative bacteria, is responsible for triggering inflammatory response during sepsis. LPS released into the circulation following bacterial lysis stimulates macrophages and endothelial cells, which eventually initiate the release of potent inflammatory mediators such as TNF-α and free radicals [19,20]. This release, in turn, activates a second level of inflammatory cascades which include other cytokines, lipid mediators, reactive oxygen species, and upregulated cell adhesion molecules which promote the migration of inflammatory cells into tissues. Any agents that are capable of neutralizing the toxic effects of LPS can ameliorate the inflammatory host response.

Cathelicidin is one novel group of these agents that can bind to LPS and hence reduce its toxicity. The less conserved C-terminal domain of cathelicidin was reported to have anti-LPS function. Human and guinea pig cathelcidins are able to suppress the binding of LPS to a CD14$^+$ murine macrophage cell line, RAW264.7, in a dose-dependent fashion. The LPS-induced TNF-α expression by the same cell line is therefore completely inhibited at both mRNA and protein level. The human cathelcidin has been further demonstrated to reduce the ability of bacterial products like LPS and lipoteichoic acid to stimulate the production of TNF-α by a human lung epithelial cell line [21]. The neutralization of LPS by cathelicidin decreases the level of a potent proinflammatory cytokine, TNF-α, and therefore prevents the mediation of inflammation.

2.2.3   Modulatory Influences in Inflammation

Inflammation is characterized by an influx of inflammatory cells into the inflamed area. Recruitment and activation of these cells are known to be regulated mainly by inflammatory cytokines such as chemokines. The upregulation of cathelicidin, during infection and inflammation [22,23] has implicated its modulatory role during inflammation which is regulated in large part by inflammatory cytokine-like chemokines, and various leukocytes are involved in the maintenance and termination. Of interest, growing evidence indicates an additional role of cathelicidin in inflammation through chemoattracting some responsible cells or altering the expression of cytokines.

To date, cathelicidin from four different mammals including bovine [24], porcine [25], human [4,26], and mouse [27] have been reported to function as leukocyte chemoattractants. They chemoattract neutrophils, monocytes, mast cells, and T-cells to the sites of inflammation at sub-antimicrobial concentrations [25]. These cells can be further stimulated to maintain inflammation through the action of adaptive cascade lymphocytes, including T cells, B cells, and antibodies. The chemotactic activity of human and mouse cathelicidins are mediated by engagement with the formyl peptide receptor-like 1 (FPRL1) receptor, a G protein-coupled, seven-transmembrane cell receptor found on macrophages, neutrophils, and subsets of lymphocytes [26,27].

Cathelicidin is also able to recruit cells by regulating the release of some chemoattractants. The human cathelicidin (LL-37) induces the secretion of

IL-8 by airway epithelial cells and hence may result in upregulated infiltration of neutrophils as well as an amplification signal of inflammation. This activation of epithelial cells has been suggested to be mediated through transactivation of the epidermal growth factor receptor (EGFR) and activation of metalloproteinases, cleavage of membrane-anchored EGFR-ligands, and activation of EGFR by these ligands [28]. in vivo studies further suggests that this peptide may stimulate macrophages to produce chemokines that, in turn, chemoattract additional cells to the inflamed sites [21].

Apart from recruiting inflammatory cells, cathelicidin is also able to modulate inflammation by altering the expression of various cytokines. The findings by Braff et al. show that LL-37 promotes the expression of numerous inflammatory mediators by keratinocytes, an epidermal resident cell. It significantly upregulates the mRNA expression of IL-8, cyclooxygenase-2 (COX-2), pro-IL-1β, and IL-6. The proinflammatory mediators TNF-α and GM-CSF are also upregulated at the protein level [29]. The ability of LL-37 to increase proinflammatory mediators released from keratinocytes may have a critical impact on the initial phase of cutaneous inflammation.

2.2.4 Clearance of Active Inflammatory Mediators

Cathelicidin, however, is toxic to eukaryotic cells, including normal and tumor cells, but at a higher concentration compared with their bactericidal activity. Its high concentration at inflammatory sites may cause degranulation of cells [30]. This cytolytic effect of cathelicidin is inhibited in an environment with serum which reflects serum proteins binding causing inhibition [30,31]. This observation has been proposed to be crucial in a host protective mechanism of healthy cells. Cathelicidin could clear up unwanted cells present at inflammation sites like activated lymphocytes. They can promote phagocytic removal of pathogen-infected cells while minimizing undesired inflammatory response [32,33].

2.3 Cathelicidin and Wound Healing

Tissue damage with subsequent repair involves a complex mechanism including influx of inflammatory cells, clearance of debris, followed by regeneration of the disrupted tissues. Recent data indicates that cathelicidin may contribute to wound repair via their angiogenic effects and their stimulatory action on migration and proliferation of epithelial cells [34–36]. All these actions promote wound healing as a host-defensive mechanism in tissues.

2.3.1 Angiogenic Influences

Vascularization, the formation of new blood vessels, is a prerequisite of tissue repair and wound healing [37,38]. It allows resupply of oxygen and nutrients to the sites of injury. It is initiated by different factors ranging from mechanical stress and hypoxia to the presence of soluble inflammatory mediators and involves sprouting of small capillaries (angiogenesis) and the growth of preexisting vessels (arteriogenesis) [38]. Based on the observation that mice with cathelicidin deficiency display decreased wound vascularization, the role of cathelicidin in angiogenesis has been investigated. Experimental findings show that LL-37 can induce functionally important angiogenesis and arteriogenesis in the chorioallantoic membrane assay and in a rabbit model of hindlimb ischemia [35].

The angiogenic activity of LL-37 is mediated by a direct action on endothelial cells and the mechanism of action is dependent on the binding of LL-37 to FPRL1 receptor that was discovered recently to mediate cellular responses to the peptide [26]. The addition of a neutralizing antiserum to FPRL1 receptor completely blocked the angiogenic activity of LL-37 on endothelial cells. Indeed, the angiogenic role of cathelicidin has been reported since 2000. The porcine cathelicidin PR39 has been found to induce both in vivo and *in vitro* angiogenesis [39]. It accelerated *in vitro* formation of vascular structures and increased myocardial vasculature in mice through inhibition of the ubiquitin-proteasome-dependent degradation of hypoxia-inducible factor-1α protein.

2.3.2 Mitogenic and Promigratory Influences

Wound repair is a complex biological process which also involves cell migration and proliferation, and extracellular matrix deposition and remodeling. The important role of cathelicidin on wound repairing was implied by the findings that the expression of cathelicidin increases sharply upon cutaneous injuries in both human and mouse epidermis [40]. Intriguingly, LL-37 is strongly expressed in healing skin epithelium and the addition of antibodies raised against LL-37 effectively suppresses reepithelialization in a dose-dependent manner in a cultured human skin model [34]. It has also been revealed that LL-37 could induce healing of airway epithelial wounds through promoting proliferation and chemotaxis of epithelial cells through EGFR, and MAP/extracellular regulated kinase [36]. By acting as a pro-migratory factor, LL-37 induces keratinocytes migration, mediated through MMP-dependent ectodomain shedding of heparin binding-EGF-like growth factor (HB-EGF), and the subsequent transactivation of EGFR and the downstream STAT3 pathway [41]. Fig. 3 summarizes the multi-host defense functions of cathelicidin against microbial infection and promotion of innate immune responses against inflammation and tissue repair in a physiological manner.

3 THE ROLE OF CATHELICIDIN IN *H. pylori* INFECTION, GASTRIC CANCER, AND MUCOSAL REPAIR IN THE STOMACH

3.1 Cathelicidin and *H. pylori* Infection

H. pylori, a spiral-shaped, Gram-negative microaerophilic stomach bacterium, is an important pathogen that colonizes >50% of the world's population [42,43]. Chronic infection with *H. pylori* is responsible for chronic gastritis,

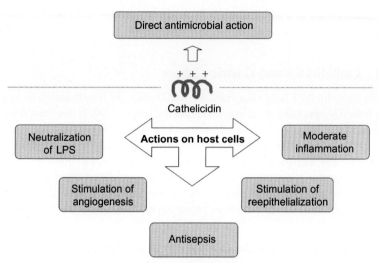

FIG. 3 Diverse biological activities of cathelicidin contributing to innate immune defense against microbial infection and inflammation and also promotion of tissue repair.

peptic ulcer, MALT lymphoma, and gastric adenocarcinoma [44,45]. As an anantimicrobial peptide, the relationship between cathelicidin expression and *H. pylori* infection was first depicted by Hase et al. [7]. They showed that LL-37 was markedly increased in the epithelium and gastric secretion of *H. pylori* infected patients, while such increase was not detected in *H. pylori*-independent inflammation. Moreover, the expression of LL-37 was reduced during progression from hyperplastic polyps, gastric adenoma, to gastric adenocarcinomas. They also showed that LL-37 was bacteriocidal against several strains of *H. pylori*. All these findings indicate that cathelicidin could play a significant role in preventing bacteria-related inflammation and perhaps also carcinogenesis in the GI tract. Later on, the effect of cathelicidin on *H. pylori* infection was further studied by the Cho's group both *in vitro* and in vivo. They found that genetic ablation of mouse cathelicidin (CRAMP) in mice significantly increased the susceptibility of *H. pylori* colonization and the associated gastritis. Replenishment with exogenous CRAMP, delivered via a bioengineered CRAMP-secreting strain of *Lactococcus lactis* (*L. lactis*), reduced *H. pylori* density in the stomach as well as the associated inflammation. Collectively, these findings indicate that cathelicidin protects against *H. pylori* infection and its associated gastritis in vivo [46]. An *in vitro* study showed that both human and mouse cathelicidin could inhibit normal and drug-resistant *H. pylori* growth, induce morphorlogical change of *H. pylori,* and human cathelicidin LL-37 could decrease biofilm formation. Furthermore, exogenous cathelicidin reversed *H. pylori* infection-induced impairment of mucus secretion and mucin gene expression [47]. Altogether, these findings indicate that cathelicidin plays a significant role as a potential natural antibiotic for *H. pylori* clearance and a therapeutic agent for chronic gastritis.

3.2 Cathelicidin and Gastric Cancer

Again study by the Cho's group also showed that LL-37 may function as a putative tumor-suppressing gene in gastric carcinogenesis [48]. It has been demonstrated that exogenous LL-37 inhibits proliferation and induces G_0/G_i-phase cell cycle arrest through activation of BMP signaling via a proteasome-dependent mechanism in gastric cancer cells. LL-37 also inhibits the growth of gastric cancer xenograft in nude mice. Moreover, depletion of endogenous LL-37 stimulates gastric cancer cell DNA synthesis suggesting that the antiproliferative effect of LL-37 occurs at physiological concentrations [48].

3.3 Cathelicidin and Wound Repair in the Stomach

As mentioned beforehand, cathelicidin could promote wound healing by stimulating angiogenesis, proliferation, and migration of endothelial cells. A pilot study by Cho's group demonstrated that rat cathelicidin can promote gastric ulcer healing through induction of cell proliferation and angiogenesis [49]. Their study showed that gastric ulceration induced the expression of rat cathelicidin (rCRAMP). Local injection of rCRAMP encoding plasmid promoted ulcer healing by enhancing cell proliferation and angiogenesis. An *in vitro* study showed that synthetic rCRAMP directly stimulated proliferation of cultured rat gastric epithelial cells through TGF-dependent transactivation of EGFR and its related signaling pathway [49].

4 THE ROLE OF CATHELICIDIN IN ULCERATIVE COLITIS AND COLON CANCER

4.1 General Introduction to Ulcerative Colitis

Ulcerative colitis (UC) and Crohn's disease (CD) are two major forms of idiopathic inflammatory bowel diseases (IBDs). They are chronic inflammatory disorders of the GI tract which have been defined empirically by their typical, clinical, pathologic, endoscopic, and laboratory features. Although the two diseases share important common pathophysiologic processes, they are classified by the region suffering from inflammation. UC is a relapsing nontransmural inflammatory disease restricted to the mucosa and superficial submucosa of the colon, whereas CD is a relapsing transmural inflammatory disease of the GI mucosa which can affect the entire GI tract from the mouth to the anus [50].

UC is the major type of IBD. However, the etiology is not yet well defined mainly because it is complicated with its causes. Indeed, there are numerous pathogenic factors contributing to the formation of UC. However, abnormal bacterial infection and immunological reaction have been reported to be the major culprits for the disease. To date there is no single drug that could have pharmacological actions against all these pathogenic factors, that is, to combat microbial infection and at the same time abrogate inflammation.

UC patients can be classified as having proctitis, left-sided colitis (involving the sigmoid colon with or without involvement of the descending colon), or pancolitis (involving the entire colon), according to the anatomic extent of involvement. A minority of the patients also develops ileal inflammation which occasionally complicates differentiation from CD. Typically, patients present with diarrhea, fever, and abdominal cramping during bowel movements [51]. Recurrence of active disease after presenting the illness has been a common characteristic. Almost all patients had at least one relapse during a 10-year period [52]. Overall, patients with UC usually have a normal life expectancy [53].

4.2   Intestinal Microbiota and Ulcerative Colitis

The pathogenesis of UC is multifactorial. It is largely caused by abnormal immunological responses and microbial affliction in the colon. The GI tract is a complex ecosystem that associates a resident microbiota and cells of various phenotypes lining the epithelial wall. Mammals are born bacteria-free but the colonization of GI tract starts immediately at birth. The resident microbiota in the GI tract is a heterogeneous microbial ecosystem containing up to 1×10^{14} colony-forming units (cfu) of bacteria [54–57]. The stomach has few resident microorganisms because of the acidic composition of the luminal medium which kills most ingested microorganism. However, *lactobacilli* have been isolated in this organ [58] in addition to the existence of *H. pylori* which is largely pathogenic in nature closely associated with the formation of gastritis and stomach cancer in humans [59,60]. In addition, the microbiota varies greatly in different parts of the GI tract. Anaerobic microorganisms are found absent in the stomach and conversely overwhelming predominant in the distal colon.

The roles of endogenous and pathogenic intestinal bacteria have been demonstrated in several animal models. Interleukin (IL)-10-deficient mice develop spontaneous colitis, while the germ-free IL-10-deficient mice remain disease free [61]. In fact, patients with IBD have higher amounts of mucosal and fecal bacteria than healthy people [62–64]. Human intestines hold about 300–500 different species of bacteria, and the number of microbial cells within the gut lumen is about 10 times larger than the number of eukaryotic cells in the human body. Studies have shown that anaerobic bacteria outnumber aerobic bacteria by a factor of 100–1000 in the GI tract. The general *Bacteroides, Bifidobacterium, Eubacterium, Clostridium, Peptococcus, Peptostreptococcus,* and *Ruminococcus* are predominant in human beings. Normal gut flora have metabolic, trophic, and protective functions. They could ferment nondigestible dietary residue and endogenous mucus, control epithelial cell proliferation and differentiation, and protect the host against pathogens. However, some of these bacteria are potential pathogens. Patients with IBD have higher amounts of bacteria attached to their epithelial surfaces than healthy people. The commensal microflora may invade the mucosa after colitis induction [65]. There is considerable indirect evidence that components of the complex microecology of the distal ileum and colon contribute to the pathogenesis of UC. Increased concentrations of *Enterobacteriaceae* and *Bacteroidaceae* species adhere to the mucosa of patients with UC and invade the mucosa, especially adjacent to ulcers and fistulae. It may also be due to the loss of tolerance toward these harmless components of the normal intestinal flora [66].

It is well established that the interaction between the intestinal flora and the mucosal defense system has a role in initiating IBD and impairment of healing [67]. Thus agents either produced endogenously in the colon or were given exogenously by the oral/anal route which can modulate these pathological activities; and/or survival of such pathogenic bacteria in the colon would have a beneficial effect against UC. There are a number of host defensive peptides, such as cathelicidin and defensin, that would have such beneficial effects [68,69].

4.3   Cathelicidin and Ulcerative Colitis

Cathelicidin expression is altered in IBD patients. It was increased in both inflamed and non-inflamed mucosa in UC patients but not in CD patients. The distribution of cathelicidin was also changed. Cathelicidin mainly expresses in the upper crypt of the colon in healthy people in contrast to the basal part in IBD patients. Elevated cathelicidin mRNA expression suggested an increase of cathelicidin production while the mucosal cathelicidin staining pattern proposed an enhancement of secretion or a translational defect in UC patients [70]. Koon et al. found that the expression of cathelicidin increased in the inflamed colonic mucosa of DSS-induced colitis and was localized to mucosal macrophages. The increase of cathelicidin may involve the activation of toll-like receptor-9-extracellular signal-regulated kinases (TLR-9-ERK) signaling [71]. It has been discovered that serum cathelicidin negatively correlates with clinical disease activity of UC patients and circulating LL-37 levels could be utilized to predict future clinical activity of UC patients [72].

We previously demonstrated that mouse cathelicidin (mCRAMP) given intrarectally could ameliorate colitis and deficiency of such peptide exacerbated inflammatory responses in mouse colons [73]. The treatment could improve different clinical symptoms including the drop of body weight, increase of fecal microflora, and relief of clinical signs in dextran sulfate sodium (DSS)-induced UC in mice. The mucus layer was also persevered through the upregulation of mucin gene and suppression of apoptosis [74]. All these actions could strengthen the mucosal defensive mechanisms in the colonic mucosa against any challenges coming from the external sources in the lumen or produced endogenously inside the mucosa [68].

UC patients are more susceptible to the challenge of luminal toxins and microorganisms due to the loss of mucus layer. The increased apoptosis in colitis could lead to the breakdown of epithelial barrier function, and thus further facilitate the mucosal invasion of intraluminal substances. LL-37 could also stimulate mucus synthesis directly through activation of *MUC1* and *MUC2* expressions and mitogen-activated protein (MAP) kinase pathway [75]. Furthermore, incorporating mCRAMP into a probiotic produces much better therapeutic outcome than the traditional antiinflammatory agent like sulfasalazine in the treatment of UC in animals. In this regard, the transformed mCRAMP-encoding *L. lactis* could effectively reduce apoptosis, maintain crypt integrity, and preserve mucus content in colons. All these actions could promote mucosal remodeling. The enhancement of mucosal repair allows for more rapid recovery of epithelial function leading to reduced exposure to various luminal agents that contribute to persistent colitis [76]. In contrast, mCRAMP knockout mice ($Cnlp^{-/-}$) carried more severe symptoms and mucosal disruption when compared with the wild-type mice in response to DSS challenge [77]. Indeed dysregulation of cathelicidin expression/function may have important roles in the pathogenesis UC [78,79].

The data reported so far demonstrates the intricate functions of cathelicidin in UC and further put forward the strategies and opportunities for cathelicidin to be a new therapeutic option and effective alternative for the prevention and treatment of UC in humans [76,80,81].

4.4   Cathelicidin and Colon Cancer

The involvement of cathelicidin in colon carcinogenesis has been reported in recent years. It was first shown by Kuroda et al. that an analogue of the LL-37 peptidedecreased the proliferation of colon cancer cells by inducing apoptosis [82]. Ren et al. from Cho's group further delineated that LL-37 was downregulated in colon cancer tissues and its expression correlated with apoptosis. They showed that LL-37 induced apoptosis in colon cancer cells through a distinct caspase-independent pathway and cathelicidin-deficient mice are more susceptible to colon cancer [83]. They also showed that a fragment of LL-37 induced concurrent activation of caspase-independent apoptosis and autophagy in colon cancer [84]. However, an in vivo study by Cheng et al. demonstrated that intravenous administration of cathelicidin expressing adeno-associated virus (AAV) or synthetic mCRAMP given through enema suppressed colon cancer without inducing apoptosis. It acts through indirect mechanism by cytoskeleton disruption, inhibition of epithelial-mesenchymal transition (EMT), and fibroblast-supported colon cancer cell proliferation [85]. In contrast to their studies, Li et al. showed that LL-37 secreted by macrophages stimulated colon cancer cell growth and neutralization of cathelicidin reduced colon tumor growth in vivo [86]. The discrepancy in these studies maybe due to the different doses of LL-37 used because in Li's study it was shown that low-dose LL-37 stimulated colon cancer growth while higher doses of LL-37 inhibited cell growth.

5 FINAL CONCLUSION USING CATHELICIDIN AS A POTENTIAL THERAPEUTIC AGENT FOR GI DISORDERS

Cathelicidins are a family of antimicrobial molecules of the innate immune system with pleiotropic activity in the regulation of inflammation and cancer of the GI tract. Cathelicidin and its recombinant protein given through an effective delivery system, like the cathelicidin-encoded *L. lactis*, could be an effective and safe treatment option of inflammation and cancer and also promotion of mucosal repair in the GI tract, especially for gastritis and UC. Indeed, more research is needed to further define the therapeutic actions and the right dosages of this host defence peptide against disorders in the GI tract. This chapter lays down the foundation for the clinical applications of cathelicidin and perhaps also other host defense peptides in the treatment of GI diseases in man.

REFERENCES

[1] Kopitar M, Ritonja A, Popovic T, et al. A new type of low-molecular mass cysteine proteinase inhibitor from pig leukocytes. Biol Chem Hoppe Seyler 1989;370:1145–51.

[2] Scocchi M, Skerlavaj B, Romeo D, et al. Proteolytic cleavage by neutrophil elastase converts inactive storage proforms to antibacterial bactenecins. Eur J Biochem 1992;209:589–95.

[3] Bals R, Wang X, Zasloff M, et al. The peptide antibiotic LL-37/hCAP-18 is expressed in epithelia of the human lung where it has broad antimicrobial activity at the airway surface. Proc Natl Acad Sci U S A 1998;95:9541–6.

[4] Agerberth B, Charo J, Werr J, et al. The human antimicrobial and chemotactic peptides LL-37 and alpha-defensins are expressed by specific lymphocyte and monocyte populations. Blood 2000;96:3086–93.

[5] Gudmundsson GH, Agerberth B, Odeberg J, et al. The human gene FALL39 and processing of the cathelin precursor to the antibacterial peptide LL-37 in granulocytes. Eur J Biochem 1996;238:325–32.

[6] Zanetti M, Litteri L, Griffiths G, et al. Stimulus-induced maturation of probactenecins, precursors of neutrophil antimicrobial polypeptides. J Immunol 1991;146:4295–300.

[7] Hase K, Murakami M, Iimura M, et al. Expression of LL-37 by human gastric epithelial cells as a potential host defense mechanism against Helicobacter pylori. Gastroenterology 2003;125:1613–25.

[8] Hase K, Eckmann L, Leopard JD, et al. Cell differentiation is a key determinant of cathelicidin LL-37/human cationic antimicrobial protein 18 expression by human colon epithelium. Infect Immun 2002;70:953–63.

[9] Travis SM, Anderson NN, Forsyth WR, et al. Bactericidal activity of mammalian cathelicidin-derived peptides. Infect Immun 2000;68:2748–55.

[10] Giacometti A, Cirioni O, Barchiesi F, et al. In vitro anti-cryptosporidial activity of cationic peptides alone and in combination with inhibitors of ion transport systems. J Antimicrob Chemother 2000;45:651–4.

[11] Giacometti A, Cirioni O, Barchiesi F, et al. In-vitro activity of polycationic peptides against *Cryptosporidium parvum*, Pneumocystis carinii and yeast clinical isolates. J Antimicrob Chemother 1999;44:403–6.

[12] Skerlavaj B, Benincasa M, Risso A, et al. SMAP-29: a potent antibacterial and antifungal peptide from sheep leukocytes. FEBS Lett 1999;463:58–62.

[13] Ahmad I, Perkins WR, Lupan DM, et al. Liposomal entrapment of the neutrophil-derived peptide indolicidin endows it with in vivo antifungal activity. Biochim Biophys Acta 1995;1237:109–14.

[14] Tamamura H, Murakami T, Horiuchi S, et al. Synthesis of protegrin-related peptides and their antibacterial and anti-human immunodeficiency virus activity. Chem Pharm Bull(Tokyo) 1995;43:853–8.

[15] Gennaro R, Zanetti M. Structural features and biological activities of the cathelicidin-derived antimicrobial peptides. Biopolymers 2000;55:31–49.

[16] Gallo RL, Nizet V. Endogenous production of antimicrobial peptides in innate immunity and human disease. Curr Allergy Asthma Rep 2003;3:402–9.

[17] Oren Z, Shai Y. Mode of action of linear amphipathic alpha-helical antimicrobial peptides. Biopolymers 1998;47:451–63.

[18] Boman HG, Agerberth B, Boman A. Mechanisms of action on Escherichia coli of cecropin P1 and PR-39, two antibacterial peptides from pig intestine. Infect Immun 1993;61:2978–84.

[19] Morrison DC, Ryan JL. Endotoxins and disease mechanisms. Annu Rev Med 1987;38:417–32.

[20] Beutler B, Cerami A. Tumor necrosis, cachexia, shock, and inflammation: a common mediator. Annu Rev Biochem 1988;57:505–18.

[21] Scott MG, Davidson DJ, Gold MR, et al. The human antimicrobial peptide LL-37 is a multifunctional modulator of innate immune responses. J Immunol 2002;169:3883–91.

[22] Frohm M, Agerberth B, Ahangari G, et al. The expression of the gene coding for the antibacterial peptide LL-37 is induced in human keratinocytes during inflammatory disorders. J Biol Chem 1997;272:15258–63.

[23] Kim ST, Cha HE, Kim DY, et al. Antimicrobial peptide LL-37 is upregulated in chronic nasal inflammatory disease. Acta Otolaryngol 2003;123:81–5.

[24] Verbanac D, Zanetti M, Romeo D. Chemotactic and protease-inhibiting activities of antibiotic peptide precursors. FEBS Lett 1993;317:255–8.

[25] Huang HJ, Ross CR, Blecha F. Chemoattractant properties of PR-39, a neutrophil antibacterial peptide. J Leukoc Biol 1997;61:624–9.

[26] De Y, Chen Q, Schmidt AP, et al. LL-37, the neutrophil granule- and epithelial cell-derived cathelicidin, utilizes formyl peptide receptor-like 1 (FPRL1) as a receptor to chemoattract human peripheral blood neutrophils, monocytes, and T cells. J Exp Med 2000;192:1069–74.

[27] Kurosaka K, Chen Q, Yarovinsky F, et al. Mouse cathelin-related antimicrobial peptide chemoattracts leukocytes using formyl peptide receptor-like 1/mouse formyl peptide receptor-like 2 as the receptor and acts as an immune adjuvant. J Immunol 2005;174:6257–65.

[28] Tjabringa GS, Aarbiou J, Ninaber DK, et al. The antimicrobial peptide LL-37 activates innate immunity at the airway epithelial surface by transactivation of the epidermal growth factor receptor. J Immunol 2003;171:6690–6.

[29] Braff MH, Hawkins MA, Di Nardo A, et al. Structure-function relationships among human cathelicidin peptides: dissociation of antimicrobial properties from host immunostimulatory activities. J Immunol 2005;174:4271–8.

[30] Johansson J, Gudmundsson GH, Rottenberg ME, et al. Conformation-dependent antibacterial activity of the naturally occurring human peptide LL-37. J Biol Chem 1998;273:3718–24.

[31] Skerlavaj B, Gennaro R, Bagella L, et al. Biological characterization of two novel cathelicidin-derived peptides and identification of structural requirements for their antimicrobial and cell lytic activities. J Biol Chem 1996;271:28375–81.

[32] Osborne BA. Apoptosis and the maintenance of homoeostasis in the immune system. Curr Opin Immunol 1996;8:245–54.

[33] Risso A, Zanetti M, Gennaro R. Cytotoxicity and apoptosis mediated by two peptides of innate immunity. Cell Immunol 1998;189:107–15.

[34] Heilborn JD, Nilsson MF, Kratz G, et al. The cathelicidin anti-microbial peptide LL-37 is involved in re-epithelialization of human skin wounds and is lacking in chronic ulcer epithelium. J Invest Dermatol 2003;120:379–89.

[35] Koczulla R, von Degenfeld G, Kupatt C, et al. An angiogenic role for the human peptide anti-biotic LL-37/hCAP-18. J Clin Invest 2003;111:1665–72.

[36] Shaykhiev R, Beisswenger C, Kandler K, et al. Human endogenous antibiotic LL-37 stimulates airway epithelial cell proliferation and wound closure. Am J Physiol Lung Cell Mol Physiol 2005;289:L842–848. Epub 2005 Jun 2017.

[37] Buschmann I, Schaper W. The pathophysiology of the collateral circulation (arteriogenesis). J Pathol 2000;190:338–42.

[38] Carmeliet P. Mechanisms of angiogenesis and arteriogenesis. Nat Med 2000;6:389–95.

[39] Li J, Post M, Volk R, et al. PR39, a peptide regulator of angiogenesis. Nat Med 2000;6:49–55.

[40] Dorschner RA, Pestonjamasp VK, Tamakuwala S, et al. Cutaneous injury induces the release of cathelicidin anti-microbial peptides active against group A Streptococcus. J Invest Dermatol 2001;117:91–7.

[41] Tokumaru S, Sayama K, Shirakata Y, et al. Induction of keratinocyte migration via transactivation of the epidermal growth factor receptor by the antimicrobial peptide LL-37. J Immunol 2005;175:4662–8.

[42] Warren JR, Marshall B. Unidentified curved bacilli on gastric epithelium in active chronic gastritis. Lancet 1983;1:1273–5.

[43] Everhart JE. Recent developments in the epidemiology of Helicobacter pylori. Gastroenterol Clin North Am 2000;29:559–78.

[44] Cover TL, Blaser MJ. Helicobacter pylori infection, a paradigm for chronic mucosal inflammation: pathogenesis and implications for eradication and prevention. Adv Intern Med 1996;41:85–117.

[45] Suerbaum S, Josenhans C. Helicobacter pylori evolution and phenotypic diversification in a changing host. Nat Rev Microbiol 2007;5:441–52.

[46] Zhang L, Yu J, Wong CC, et al. Cathelicidin protects against Helicobacter pylori colonization and the associated gastritis in mice. Gene Ther 2013;20:751–60.

[47] Zhang L, Wu WK, Gallo RL, et al. Critical role of antimicrobial peptide cathelicidin for controlling Helicobacter pylori survival and infection. J Immunol 2016;196:1799–809.

[48] Wu WK, Sung JJ, To KF, et al. The host defense peptide LL-37 activates the tumor-suppressing bone morphogenetic protein signaling via inhibition of proteasome in gastric cancer cells. J Cell Physiol 2010;223:178–86.

[49] Yang YH, Wu WK, Tai EK, et al. The cationic host defense peptide rCRAMP promotes gastric ulcer healing in rats. J Pharmacol Exp Ther 2006;318:547–54.

[50] Baumgart DC, Sandborn WJ. Inflammatory bowel disease: clinical aspects and established and evolving therapies. Lancet 2007;369:1641–57.

[51] Podolsky DK. Inflammatory bowel disease. N Engl J Med 1991;325:928–37.

[52] Hendriksen C, Kreiner S, Binder V. Long term prognosis in ulcerative colitis-based on results from a regional patient group from the county of Copenhagen. Gut 1985;26:158–63.

[53] Winther KV, Jess T, Langholz E, et al. Survival and cause-specific mortality in ulcerative colitis: follow-up of a population-based cohort in Copenhagen County. Gastroenterology 2003;125:1576–82.

[54] Vaughan EE, Schut F, Heilig HG, et al. A molecular view of the intestinal ecosystem. Curr Issues Intest Microbiol 2000;1:1–12.

[55] Morelli L, Cesena C, de Haen C, et al. Taxonomic lactobacillus composition of feces from human newborns during the first few days. Microb Ecol 1998;35:205–12.

[56] Berg RD. The indigenous gastrointestinal microflora. Trends Microbiol 1996;4:430–5.

[57] Zboril V. Physiology of microflora in the digestive tract. Vnitr Lek 2002;48:17–21.

[58] Roach S, Savage DC, Tannock GW. Lactobacilli isolated from the stomach of conventional mice. Appl Environ Microbiol 1977;33:1197–203.

[59] Cover TL, Blaser MJ. Helicobacter pylori and gastroduodenal disease. Annu Rev Med 1992;43:135–45.

[60] Parsonnet J, Friedman GD, Vandersteen DP, et al. Helicobacter pylori infection and the risk of gastric carcinoma. N Engl J Med 1991;325:1127–31.

[61] Sellon RK, Tonkonogy S, Schultz M, et al. Resident enteric bacteria are necessary for development of spontaneous colitis and immune system activation in interleukin-10-deficient mice. Infect Immun 1998;66:5224–31.

[62] Swidsinski A, Ladhoff A, Pernthaler A, et al. Mucosal flora in inflammatory bowel disease. Gastroenterology 2002;122:44–54.

[63] Sokol H, Seksik P, Rigottier-Gois L, et al. Specificities of the fecal microbiota in inflammatory bowel disease. Inflamm Bowel Dis 2006;12:106–11.

[64] Dickinson RJ, Varian SA, Axon AT, et al. Increased incidence of faecal coliforms with in vitro adhesive and invasive properties in patients with ulcerative colitis. Gut 1980;21:787–92.

[65] Guarner F, Malagelada JR. Gut flora in health and disease. Lancet 2003;361:512–9.

[66] Duchmann R, Kaiser I, Hermann E, et al. Tolerance exists towards resident intestinal flora but is broken in active inflammatory bowel disease (IBD). Clin Exp Immunol 1995;102:448–55.

[67] Kucharzik T, Maaser C, Lugering A, et al. Recent understanding of IBD pathogenesis: implications for future therapies. Inflamm Bowel Dis 2006;12:1068–83.

[68] Chow JY, Li ZJ, Wu WK, et al. Cathelicidin a potential therapeutic peptide for gastrointestinal inflammation and cancer. World J Gastroenterol 2013;19:2731–5.

[69] Haney EF, Mansour SC, Hancock RE. Antimicrobial peptides: an introduction. Methods Mol Biol 2017;1548:3–22.

[70] Schauber J, Rieger D, Weiler F, et al. Heterogeneous expression of human cathelicidin hCAP18/LL-37 in inflammatory bowel diseases. Eur J Gastroenterol Hepatol 2006;18:615–21.

[71] Koon HW, Shih DQ, Chen J, et al. Cathelicidin signaling via the Toll-like receptor protects against colitis in mice. Gastroenterology 2011;141:1852–63. e1851–1853.

[72] Tran DH, Wang J, Ha C, et al. Circulating cathelicidin levels correlate with mucosal disease activity in ulcerative colitis, risk of intestinal stricture in Crohn's disease, and clinical prognosis in inflammatory bowel disease. BMC Gastroenterol 2017;17:63.

[73] Tai EK, Wu WK, Wang XJ, et al. Intrarectal administration of mCRAMP-encoding plasmid reverses exacerbated colitis in Cnlp(−/−) mice. Gene Ther 2013;20:187–93.

[74] Tai EK, Wu WK, Wong HP, et al. A new role for cathelicidin in ulcerative colitis in mice. Exp Biol Med (Maywood) 2007;232:799–808.

[75] Tai EK, Wong HP, Lam EK, et al. Cathelicidin stimulates colonic mucus synthesis by up-regulating MUC1 and MUC2 expression through a mitogen-activated protein kinase pathway. J Cell Biochem 2008;104:251–8.

[76] Wong CC, Zhang L, Wu WK, et al. Cathelicidin-encoding Lactococcus lactis promotes mucosal repair in murine experimental colitis. J Gastroenterol Hepatol 2017;32:609–19.

[77] Tai EK, Wu WK, Wang XJ, et al. Intrarectal administration of mCRAMP-encoding plasmid reverses exacerbated colitis in Cnlp(−/−) mice. Gene Ther 2012;20:187–93.

[78] Wehkamp J, Schmid M, Stange EF. Defensins and other antimicrobial peptides in inflammatory bowel disease. Curr Opin Gastroenterol 2007;23:370–8.

[79] Wong CC, Zhang L, Ren SX, et al. Antibacterial peptides and gastrointestinal diseases. Curr Pharm Des 2011;17:1583–6.

[80] Ahluwalia A, Tarnawski AS. Cathelicidin gene therapy: a new therapeutic option in ulcerative colitis and beyond? Gene Ther 2013;20:119–20.

[81] Sun L, Wang W, Xiao W, et al. The roles of cathelicidin LL-37 in inflammatory bowel disease. Inflamm Bowel Dis 2016;22:1986–91.

[82] Kuroda K, Fukuda T, Yoneyama H, et al. Anti-proliferative effect of an analogue of the LL-37 peptide in the colon cancer derived cell line HCT116 p53+/+ and p53. Oncol Rep 2012;28:829–34.

[83] Ren SX, Cheng AS, To KF, et al. Host immune defense peptide LL-37 activates caspase-independent apoptosis and suppresses colon cancer. Cancer Res 2012;72:6512–23.

[84] Ren SX, Shen J, Cheng AS, et al. FK-16 derived from the anticancer peptide LL-37 induces caspase-independent apoptosis and autophagic cell death in colon cancer cells. PLoS One 2013;8.

[85] Cheng M, Ho S, Yoo JH, et al. Cathelicidin suppresses colon cancer development by inhibition of cancer associated fibroblasts. Clin Exp Gastroenterol 2015;8:13–29.

[86] Li D, Liu W, Wang X, et al. Cathelicidin, an antimicrobial peptide produced by macrophages, promotes colon cancer by activating the Wnt/beta-catenin pathway. Oncotarget 2015;6:2939–50.

Chapter 5

Antimicrobial Peptides as Potential Therapy for Gastrointestinal Cancers: Opportunities and Challenges

Li Ma* and James P. Dilger†
*Department of Science Education, Donald and Barbara Zucker School of Medicine at Hofstra/ Northwell, Hempstead, NY, United States, †Department of Anesthesiology, Stony Book Medicine, Stony Brook University, Stony Brook, NY, United States

Chapter Outline

Antimicrobial Peptides in Gastrointestinal Diseases. https://doi.org/10.1016/B978-0-12-814319-3.00005-2
Copyright © 2018 Chi Hin Cho. Published by Elsevier Ltd. All rights reserved.

1 INTRODUCTION

Two classes of antimicrobial peptides (AMPs), defensins and cathelicidins, were first identified in humans in the mid-1980s and mid-1990s, respectively. These AMPs are secreted by immune cells and by epithelial cells from, for example, the respiratory, urogenital, and gastrointestinal (GI) tracts. Most defensins and cathelicidins are constitutively expressed and can also be induced under pathological conditions, such as infection, inflammation, and injury. In retrospect, these peptides could have been named "host defense molecules." In addition to their antimicrobial activity, AMPs exhibit antiviral, antifungal, wound healing, and immunoregulatory effects [1–5].

Human cathelicidins and defensins also affect tumors. An increasing amount of evidence suggests that AMPs have both stimulating and inhibitory effects on tumor growth [6]. The mechanisms are tissue-specific, complex, and appear to depend on peptide concentration. Both overexpression and down-regulation of cathelicidin and defensins have been observed in cancer tissues. However, it is not known whether the changes in regulation are due to changes in constitutive AMPs from epithelial cells, changes in inducible AMPs from tumor stroma cells, or both. Research in this field is developing rapidly, yet there are still more questions than answers.

In this chapter, we will critically review literatures on how human cathelicidin and defensins affect tumorigenesis and progression, particularly in the GI tract. We will examine the proposed mechanisms of action, potential clinical uses of exogenous AMPs, and the challenges that still must be faced.

2 STRUCTURE AND CLASSIFICATION OF HUMAN AMPs

Structurally, cathelicidins are linear α-helical peptides, whereas, defensins are β-strand peptides connected by disulfide bonds [7,8].

There are two main subfamilies of human defensins: α and β defensins. They differ in the length of the peptide segments between the six cysteines and in the pairing of the cysteines that are connected by disulfide bonds.

Human α-defensins are arginine-rich peptides, containing 29–35 amino acids, including six cysteine residues that create three disulfide bonds (1–6, 2–4, 3–5). The six known forms of human α-defensins are distinguished by the types of cells that secrete them. Human Neutrophil Proteins 1 through 4 (HNP1–4) are expressed in immune cells such as neutrophils and lymphocytes. Human Defensins 5 and 6 (HD5 and HD6) are primarily produced by Paneth cells in the crypts of the small intestine and the epithelial cells of the female urogenital tract [7,8].

Human β-defensins (hBDs) are peptides of about 35 amino acid residues, including six cysteine residues that create three disulfide bonds (1–5, 2–4, 3–6). They are expressed predominantly in epithelial tissues, such as skin, and gastrointestinal, respiratory, and urinary tracts. However, hBD3 has also been found expressed in monocytes [9]). hBDs provide the first line of defense between humans and their environment. More than 50 hBD genes have been identified in the human genome so far, among them, hBD1, hBD2, hBD3, and hBD4 are best characterized [7,8].

Human cathelicidin antimicrobial protein, hCAP18, is the only member of the cathelicidin family found in human cells. Its mature peptide, the C-terminal domain released by proteolytic cleavage, is referred to as LL-37 because it begins with two leucine residues and is 37 amino acid residues long [10]. Similar to human defensins, cathelicidin peptides are expressed in various types of cells including skin, epithelial cells of the intestine, respiratory system, genitals, ocular surface, differentiated surface and upper crypt epithelial cells in the colon, Brunner glands in the duodenum and in eccrine glands, as well as immune cells [11].

3 EXPRESSION OF CATHELICIDIN AND DEFENSINS IN GI CANCER

3.1 LL-37

LL-37/hCAP18 messenger RNA and protein are expressed in mucous-producing surface epithelial cells, epithelial cells in the upper gastric pits, and fundic glands located in the basal region. Epithelial LL-37/hCAP18 expression is significantly decreased in gastric hyperplastic polyps, tubular adenomas, and adenocarcinomas [12]. The issue of how expression of LL-37 is regulated in colon cancer is complex and controversial. Ren and coworkers studied colonic tissue from cancer patients and noncancerous control patients. In both noncancerous tissue from cancer patients and tissue from patients without cancer, the percentage of cells expressing LL-37 varied widely from 0% to 90%. In contrast, LL-37 expression was observed in 0%–30% of colon cancer cells; a highly significant difference. Moreover, a comparison of the cancer cells with the adjacent noncancerous cells in the same patient revealed that in the majority of patients, LL-37 expression in the cancer tissue was decreased to <10% of the cells. They also detected the presence of LL-37 in submucosal leukocytes from the cancerous tissues [2]. Li and coworkers studied 68 human colon cancer tissue sections and 30 noncancerous tissues. They found very little evidence for LL-37 in epithelial cells in both cancerous and noncancerous epithelial cells. However, infiltrating inflammatory immune cells in the cancerous tissue did express high levels of LL-37. They also found that the serum concentration of LL-37 was much higher in patients with colon cancer compared to healthy humans, and that these levels decreased after surgical removal of colon tumors [13]. The two studies seem to agree that when LL-37 is detected in colon cancer tissues, it arises mostly from the immune cells rather than the epithelial cells in the tissue.

3.2 Beta Defensins

hBDs 1–4 are epithelium-derived AMPs found on the surface of mucosa that are essential for the host defense response. hBD-1 is thought to be the only β defensin that is constitutively expressed in the normal GI mucosa. Expression of the other β defensins is induced in these tissues only in response to infection, inflammation, or some other stimulus [7,14].

An immunohistochemical study of hBD-3 in surgically resected colon cancer tissues revealed that the colon cancer cells themselves did not express hBD-3; however, many tissue samples stained positive for hBD-3 due to infiltrating monocytes surrounding the cancer cells [9]. This finding suggests that great care must be taken to distinguish among the different cell types in tissue samples collected from humans. Immunohistochemical and qPCR evidence for differences in hBD expression between cancerous and noncancerous samples of colonic tissue [15] may have to be reassessed in terms of cell type.

3.3 Alpha Defensins HNP 1–3

Human neutrophil α-defensins 1–3 are upregulated in the microenvironment of colon and gastric tumors compared to that of nearby normal tissue. Because these AMPs are primarily expressed by immune cells rather than by enteric cells themselves, this upregulation presumably represents a selective accumulation of immune cells in the cancerous tissue. However, there is some evidence to suggest that some HNPs are expressed in cancerous tissue (see later).

Although patient-to-patient variability is high, the amount of HNP1–3, determined by mass spectrometry, was seen to be larger in tumors compared to normal colonic tissue in colorectal cancer patients in Dukes' stage A–D [16,17]. Upregulation of HNP 1–3 mRNA expression was also observed in colon adenoma tissue, carcinoma tumor tissue, and the transition area between normal and tumor tissues compared to normal samples [18]. Because the samples used in these studies were mixtures of tumor and tumor stroma cells, they were unable to identify the source of the upregulated HNP1–3. Melle and coworkers reported similar results that both infiltrating inflammatory cells and epithelial tumor cells contributed to the upregulation of HNPs [19]. Pagnini and coworkers also presented convincing immunohistological evidence for HNP-1 being expressed within both neutrophils and the epithelial cells of colonic adenoma patients [20].

The plasma concentration of HNP1–3 as determined by ELISA was significantly elevated in colon cancer patients compared to healthy controls [19]. Subsequently, Albrethsen and coworkers showed that this was true only for patients with more advanced cancer (Dukes' stages C and D) [16,17]. The most recent study of plasma HNP 1–3 levels reported that they have high predictive values for identifying colon cancer patients. The mean levels were 3954 + 678 pg/ml in colon cancer patients compared to the levels in control volunteers 68 + 22 pg/ml, $p < 0.0001$ [21].

Several studies have shown that the expression of HNPs 1 3 in gastric cancer tissues is increased compared with normal adjacent tissues [22,23]. Using matrix-assisted laser desorption ionization time-of-flight mass spectrometry (MALDI-TOFMS), Cheng and coworkers demonstrated that levels of HNPs 1–3 were significantly higher in cancerous tissues of gastric cancer patients compared to neighboring noncancerous tissues. Immunohistology indicated that HNP-1 was highly expressed in the lamina propria of cancerous tissue. In addition, expression of HNPs showed a strong positive correlation with the number of infiltrating neutrophils [22]. Mohri and coworkers analyzed tissue using surface-enhanced laser desorption/ionization time-of-flight mass spectrometry (SELDI) and also showed that HNPs 1–3 were elevated 10-fold in gastric cancer relative to adjacent normal mucosa. Immunohistochemistry localization of the expression of HNPs 1–3 suggested that the elevated HNPs 1–3 in gastric cancers are primarily from the epithelial cells of the tumors rather than the infiltrating neutrophils. [23]. Thus far, nothing has been published concerning the serum levels of HNPs in gastric cancer patients.

Overall, the results consistently demonstrate elevation of HNP1–3 in gastric and colon tumors, tissues consisting of a mixture of epithelial cells, fibroblasts, and immune cells. It is most likely that both invasion of the neutrophils and the cancer cells themselves contribute to the upregulated HNP1–3 expression in gastric and colon cancer tissues. Changes of plasma HNP1–3 may be a potential marker for prognostic assessment, surveillance of patients, and monitoring chemotherapy in colorectal cancer patients with advanced disease [16,17].

3.4 Alpha Defensins HD5, HD6

Enteric α-defensins HD5 and HD6 are present in intestinal Paneth's cells and are also constitutively expressed in the epithelial cells of the small intestine. The levels of expression of HD5 and HD6 genes in the mucosa from adenomatous polyps are dramatically higher than what is seen in normal colon mucosa. Hyperproduction of HD5 and HD6 proteins were observed in both adenoma [18,20] and carcinoma tissues [18] compared with the normal mucosa. Immunohistology indicated that HD6 is located within the adenoma epithelial cells as expected [20]. Interestingly, the expression of HD6 in adenoma is 60-fold higher than in the carcinoma tissue. This sharp distinction between adenoma and fully blown carcinoma may be envisioned as a clinically-useful diagnostic marker [18]. In addition, the serum levels of HD5 [24,25] and HD6 [24,25] in colon cancer patients were significantly higher than those found in normal subjects.

4 THE ROLES OF CATHELICIDIN (LL-37) AND DEFENSINS IN TUMORIGENESIS

LL-37 and defensins demonstrate both pro- and anticancer activities. The mechanisms that determine whether and when these AMPs exert pro- or anticancer effects have not been fully elucidated. Originally, it was believed that the effects of cathelicidin and defensins on tumorigenesis were tissue specific. For example, they have procancer effects in ovarian, lung, breast, and skin cancers, but show anticancer effects in colon and gastric cancer [26]. Our review of the published in vitro studies suggests that the effects are not only tissue specific but may also be dose-dependent.

4.1   In Vitro Studies

4.1.1   Cathelicidin

Cathelicidin and GI Cancer

Until recently, LL-37 was univocally considered to be an anticancer peptide in colon cancer. LL-37 and its related peptides inhibit colon cancer cell proliferation [27] and induce cell apoptosis [2,28]. However, Li and coworkers have shown that LL-37 has biphasic effects on colon cancer cell proliferation, depending on its concentrations [13]. At low concentrations (50 ng-1 µg/ml), LL-37 stimulates the proliferation of HCT116 and SW480 colon cancer cells. LL-37 inhibits proliferation only at concentrations above 5 µg/ml. The earlier in vitro studies had been performed using high concentrations of either LL-37 or its related peptides ranging from 40 to 360 µg/ml [2,27,29].

Cathelicidin and Other Cancers

Studies of other types of human cancers, including lung, skin squamous cell carcinoma, malignant melanoma and ovarian, reveal a similar bell-shaped concentration-effect relationship for cell proliferation by LL-37 (Fig. 1). At low concentrations, LL-37 promotes the growth of cancer cells (albeit not as dramatically as in colon cancer cells), and at high concentrations, it inhibits growth. Some studies examined the effects of LL-37 on cancer cell migration and invasion [31,32] and found a similar bell-shaped concentration-effect relationship. We are not aware of any study showing an effect of exogenous LL-37 on breast cell proliferation. However, LL-37 at 4.5 or 9 µg/ml stimulates migration and invasion of MCF7 breast cancer cells [34,35]. Taken together, these results indicate that low concentrations of LL-37 not only promote cancer cell growth, but also potentiate the spread of the cancer cells. High concentrations of LL-37 exert anticancer activity, regardless of the cancer cell types.

4.1.2   Defensins and Cancer Cells

There is only one published in vitro study of the effects of human defensins on GI cancer cells. Incubation of gastric cancer cells (AGS) with 10, 50, or 100 µg/ml HNP-1 for 2 days demonstrated a significant concentration-dependent inhibition of cell proliferation. At the highest concentration, cell viability was reduced to 25% of normal [22].

The effects of HNP-1 over a larger range of concentrations have been studied in malignant oral epithelial cells [36]. Concentrations between 1 and

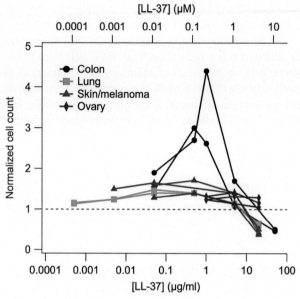

FIG. 1 LL-37 has cell- and concentration-dependent effects on the growth of cancer cell lines in vitro. Data are normalized to cell growth in the absence of LL-37. Data sources: colon cancer cells [13], lung cancer cells [30], skin squamous cell carcinoma [31], malignant melanoma cells [32], and ovarian cancer cells [33].

10 μg/ml tended to produce a moderate (up to 1.5-fold) increase in cell proliferation. Concentrations of 20–50 μg/ml inhibited cell growth. The high concentrations of HNP-1 were also cytotoxic to normal epithelial cells. The same paper examined beta-defensins (hBD-1, -2, -3) over the same concentration range. High concentrations of hBD-1 and hBD-2 showed a moderate inhibition in the squamous cell carcinoma KB cell line, but not in SCC9. Low concentrations of hBD-2 increased proliferation in both cell lines. Low concentrations of hBD-3 increased proliferation in KB cells only and high concentrations had no effect on either cell line [36].

hBD-2 is overexpressed in esophageal SCC cell lines (KYSE-70, KYSE150, KYSE-270, and KYSE-410) relative to esophageal normal cell line (HET-1A). Knockdown of hBD-2 in esophageal SCCs inhibits cell proliferation and migration in vitro [37]. The concentration of hBD-2 in the condition media was not given.

Although hBD-3 was not detected in colon cancer cells, many tissue samples stained positively for hBD-3 due to infiltrating monocytes surrounding the cancer cells. Incubation of human colon cell lines SW480 and SW620 with hBD-3 at 1 and 5 μM (~5 and 25 μg/ml) did not affect cell proliferation but significantly inhibited tumor cell migration [9].

Alpha defensins, HNPs-1, -2, and -3 are expressed not only in neutrophils, but also found in renal cell carcinomas (RCCs) in vitro and in vivo.

Concentrations of 6–25 µg/ml HNPs-1, -2, -3 strongly stimulated cell proliferation in the A-498 and 786-O RCC cell lines but not the WT-33, ACHN and 769-P RCC cell lines. Interestingly, HNPs-1, -2, -3 at higher concentration (>25 µg/ml) inhibited cell proliferation in all RCC cell lines [38].

Although the data on the effects of defensins on GI cancer cell proliferation are sparse, comparison with other types of cancers shows that both increases and decreases in proliferation can occur. The effects depend on the type of cancer cell, the identity of the defensin molecule, and the concentration of the defensin.

4.2 In Vivo Studies

4.2.1 In Vivo Studies of LL-37

Both upregulation and down regulation of LL-37 have been observed in human tumor tissues. In vivo, LL-37 can originate from both cancer cells and tumor stroma cells, including immune cells and multipotent mesenchymal stromal/ stem cells (MSCs). Here, we look both sources of LL-37 in GI cancer and compare this with what is seen in other types of cancers.

LL-37 in GI Cancer

Expression of LL-37 is decreased in human colon tumor cells [2] but increased in colon infiltrating macrophages [13]. Using a chemically induced mouse colon cancer model (azoxymethane+dextran sodium sulfate ×3), Li and coworkers found that mouse cathelicidin (cathelicidin-related AMP=CRAMP) behaved similarly. CRAMP was highly expressed within the macrophages of the stroma but not in the colon tumor cells themselves. An anti-CRAMP antibody added at an early stage of tumor development reduced the engraftment of macrophages into colon tumors, deactivated the Wnt/β-catenin signaling pathway, and inhibited proliferation of tumor cells, resulting in an inhibition of tumor growth [13]. These results would suggest that cathelicidin induced by macrophages promotes tumor growth. However, a different study showed that cathelicidin deficient (CRAMP knockout) mice were more susceptible than wild-type mice to azoxymethane-induced gross colon tumors [2]. This result would suggest that cathelicidin inhibits tumorigenesis. The discrepancy between the two animal studies may be due to the dual roles of cathelicidin in tumor initiation and progression. Constitutive expression of cathelicidin in the colon mucosa may play a surveillance role preventing tumorigenesis, whereas once a tumor starts, cathelicidin secreted by the tumor stroma cells, together with other diverse factors in the tumor microenvironment, promotes tumor progression. A more thorough understanding of these intertwined pathways is essential.

FF/CAP18 (designed by replacement of a glutamic acid and a lysine residue of LL-37 with phenylalanine which increases its antimicrobial activity) has antitumorigenesis effects in a mouse xenograft model with human colon cancer cells (HCT116). Tumor volume and final weight were reduced significantly in mice simultaneously inoculated with HCT116 cells and FF/CAP18 (10mg/kg) compared with those inoculated with HCT116 cells and scrambled control peptide (10mg/kg) [39].

In a nude mouse xenograft model with human gastric TMKI cells, local subcutaneous injections of LL-37 (40µg/mouse) were started 10days after inoculation of tumor cells. Injection of LL-37 at sites adjacent to the tumor every other day over 1week inhibited gastric tumor size by 40% [3].

LL-37 in Other Cancers

Two studies using mouse xenograft models with human breast cancer cells give conflicting results regarding the effects of LL-37. More metastasis was observed in mice injected with MCF7 cells that overexpress hCAP-18/LL-37 than in mice injected with regular MCF7 cells. This suggests a protumor role for LL-37 [35]. In contrast, when calcitriol was used to increase LL-37 expression in mice inoculated with HCC1806 human breast cancer cells, tumor growth was reduced [40]. It is not clear whether these results indicate cell-dependent and or concentration-dependent actions of LL-37.

In normal human ovarian tissue, hCAP-18/LL-37 expression is low and is limited to the immune and granulosa cells found in the organ. In ovarian tumors, higher-than-normal levels of hCAP-18/LL-37 are found in both tumor and stromal cells [33]. In addition, there is evidence that ovarian tumor-derived LL-37 recruits multipotent mesenchymal stromal/stem cells (MSCs) from bone marrow into the tumor stroma, consequently promoting tumor growth by producing high levels of LL-37 [41]. In contrast to the protumorigenesis effects of LL-37-mediated recruitment of MSCs, LL-37 has shown synergistic toxicity against ovarian cancer cells in combination with Toll-like receptor (TLR) ligands such as CpG oligodeoxynucleotides (CpG-ODN) [42]. Treatment of murine ovarian tumor-bearing mice with the combination of CpG-ODN (30µg/mouse) and LL-37 (100µg/mouse) resulted in smaller tumors and longer survival compared with treatment with CpG-ODN or LL-37 alone [42].

Primary human pancreatic ductal adenocarcinoma (PDAC) tissue expresses higher levels of hCAP-18/LL-37 compared to normal pancreas. In addition, the upregulated LL-37 is primarily restricted to the macrophages within the PDAC stroma. Wild-type mice had a higher number of cancer stem cells (CSCs) and a faster tumor formation rate than CRAMP−/− mice. Tumor growth is also faster in mice inoculated with CSC enriched murine PDAC cells plus wild type macrophages compared to that in mice inoculated with the same cancer cells plus CRAMP−/− macrophages [43].

Overall, these studies suggest that in vivo, tumor-derived LL-37 is able to recruit various types of tumor stroma cells, such as immune cells, which in turn synthesize and secrete more LL-37. Stroma cell-produced LL-37 promotes progression by stimulating tumor cell proliferation, migration, and invasion. It is likely that constitutively expressed LL-37 has antimicrobial and antiinflammation activities, which inhibit tumorigenesis. However, once tumors are established, LL-37 promotes tumor progression and metastasis. These intertwined relationships make the role of LL-37 in tumorigenesis difficult to resolve, with supporting evidence for both pro- and anticancer activities.

4.2.2 In Vivo Study of Defensins

We are not aware of any in vivo studies investigating the role of defensins in colon or gastric cancer. In a mouse xenograft of a human Lewis lung carcinoma cell model, Defb14 (the mouse homolog of hBD3), continuously infused subcutaneously near the tumor site by a MINI-Osmotic Pump for 9 days, demonstrated a significant reduction in tumor growth [44].

5 MECHANISMS OF ANTITUMOR ACTIVITY OF CATHELICIDIN AND HUMAN DEFENSINS

AMPs are generally polycationic and amphipathic. These characteristics enable them to interact with the negatively charged membrane of microorganisms and to partition into the membrane. The classic explanation for the antibacterial activity of AMPs is that after they partition into the outer and inner membranes of bacterial cells they form voltage-dependent channels that increase the permeability of the membranes to ions and other small molecules. This may result in cell lysis. Recent evidence, however, suggests after adsorbing to the cell membrane, AMPs are internalized into the cytoplasm where they may target intracellular proteins responsible for protein synthesis or cell metabolism [45].

The polycationic and amphipathic nature of AMPs has also been proposed to account for the selective interaction of AMPs with cancer cells over normal cells. Compared to normal cells, the membranes of cancer cells contain more negatively charged lipids such as phosphatidylserine, an increased number of sialic acid residues, and higher expression of heparan sulfate proteoglycans. In addition, the low membrane fluidity of cancer cells in solid tumors, the acidity of the tumor environment, as well as large cell surface area due to increased numbers of microvilli also promote the interaction of AMPs with the membrane of cancer cells [46]. However, there is no consensus about what AMPs do after they enter the cancer cell membrane. Various mechanisms have been proposed to account for the antitumor effects of cathelicidin and defensins. This includes membrane disruption [47], mitochondria damage [27,48], regulation of microRNA expression [39] and gene transcription [49], alteration of cancer cell metabolism [28], as well as changes in host immune system-mediated anticancer activities [42,50].

5.1   Cell Membrane Lysis

When HNP-1 (25–100 µg/ml) is incubated with human leukemia K562 cancer cells, it binds to the cells and induces cell lysis in a dose-dependent manner. In contrast, HNP-1 binds to a similar extent to L929 mouse fibrosarcoma cells but does not induce membrane permeabilization. This argues that interaction with cell membranes is necessary but may not be sufficient for HNP-1 to produce cytotoxicity in all tumors [47]. In addition, HNP-1 causes a collapse of the plasma membrane potential of K562 cells within minutes, but cell lysis occurs more slowly. This leads to the postulate that defensin must partition from the membrane into the cytoplasm and act through an intracellular process to induce cytotoxicity [47].

5.2   Mitochondrial Membrane Depolarization and DNA Fragmentation

As noted in Section 4.2.1, FF/CAP18 (10 and 40 µg/ml, high concentrations) reduces proliferation of the human colon cancer cell line HCT116. Dye studies indicated that FF/CAP18 depolarizes the inner mitochondrial membrane, an early step in apoptosis. FF/CAP18 also induces DNA fragmentation in a p53-independent manner [27]. Furthermore, a 27-mer peptide of the C-terminal domain (hCAP18$_{109-135}$) promotes caspase-independent mitochondrial depolarization and apoptosis in human oral squamous cell carcinoma SAS-H1 cells. However, at the same concentration, hCAP18$_{109-135}$ has no significant effect on normal human gingival fibroblasts and human keratinocyte HaCaT cells [48].

5.3   Apoptosis and Autophagy

FK-16, a fragment of LL-37, retains the antibiotic and anticancer properties of full-length LL-37. Similar to LL-37, high concentrations (45–180 µg/ml) of FK-16 activates caspase-independent (AIF/EndoG-dependent) apoptosis in human colon cancer cell lines. Unlike LL-37, FK-16 also induces cell death via LC3-dependent autophagy [51]. Their results further demonstrate that p53-mediated upregulation of Bax and downregulation of Bcl-2 are associated with both apoptosis and autophagic cell death induced by FK-16. In contrast, LL-37 induces only apoptosis but not autophagy in these colon cancer cells [51].

5.4   Regulation of MicroRNA

Recently, miR663a has been recognized as a tumor suppressing microRNA in several cancer cells including colon cancer cells [51a]. High concentrations of LL-37 upregulate miR-663a in HCT116 colon cancer cells. Overexpression of miR-663a inhibits tumor cell growth both in vivo and in vitro via suppression of C-X-C chemokine receptor type 4 (CXCR4)-Akt pathway [39]. These results suggest that upregulation of miR663a may be one of the antitumor mechanisms of LL-37.

5.5 Alteration of Tumor Cell Metabolic Profile

A recent metabolomics study determined that FF/CAP18 causes widespread changes in the metabolic profiles of HCT116 colon cancer cells. Cancer cells obtain ATP primarily from glycolysis rather than mitochondrial respiration even in the presence of oxygen. At the lower concentration of FF/CAP18 studied (10 μg/ml), the glycolytic pathway was restricted and mitochondrial respiration was upregulated. This may be responsible for the initiation of apoptosis. The higher concentration of FF/CAP18 (40 μg/ml) caused decreased ATP production by both glycolysis and the TCA cycle leading to the later stages of apoptosis [28]. It would be interesting to determine the effects of lower concentrations of FF/CAP18 on the metabolic profiles. It may be that smaller changes in metabolism might be insufficient to initiate apoptosis but, instead, might be beneficial to the survival of the cancer cells.

5.6 Activation of Bone Morphogenetic Protein (BMP)

BMPs are a group of growth factors most of which are members of the transforming growth factor β (TGF-β) super-family. BMPs control the development of bone and many other tissues. Similar to LL-37, BMP signaling in the stomach and colon is upregulated during inflammation and downregulated in cancer [52–54]. Mice that overexpress the BMP antagonist Noggin in the intestine develop intestinal tumors [55]. In addition, when BMP signaling was conditionally inactivated in transgenic mice, cancerous lesions developed at gastric epithelial transition zones [56]. LL-37 (4–40 μg/ml) was found to inhibit gastric cancer cell proliferation via activation of BMP signaling through a mechanism that involves the ubiquitin-proteasome pathway [3].

5.7 Inhibition of Cancer Cell Migration

Human colon cancer cells do not express hBD-3, but the peptide is introduced into colon cancer tumors by monocytes. hBD-3 (1–5 μg/ml) inhibited the migration but not the proliferation of colon cell cultures. Colon cancer cells exposed to hBD-3 showed decreased mRNA for metastasis-associated 1 family member 2 (MTA2), a regulator of gene expression [9].

5.8   Regulation of Gene Transcription

In melanoma cells, LL-37 was found to migrate to the nucleus and bind to gene promoter regions. Microarray analyses suggest that LL-37 modifies the transcriptional activity of genes associated with histone, metabolism, cellular stress, ubiquitination, and mitochondria [49]. Thus far, this mechanism has not been reported to operate in GI cancer.

5.9   Regulation of Host Immune System (Indirect Antitumor Activity)

Toll-like receptors (TLRs) play a major role in host immune responses. They recognize ligands such as bacterial cell-surface lipopolysaccharides and bacterial DNA. In cancer immunotherapy, CpG oligodeoxynucleotides are used to enhance tumor-suppressing activity of the host immune cells through stimulation of TLR9. LL-37 can enhance the sensing of CpG oligodeoxynucleotides by B lymphocytes, plasmacytoid dendritic cells, and natural killer (NK) cells [42,57]. Coadministration of LL-37 with CpG oligodeoxynucleotides produces synergistic antitumor effects against ovarian cancer in mice by increasing the delivery of CpG oligodeoxynucleotides into the endosomal compartments. Moreover, CpG oligodeoxynucleotides and LL-37 enhance the proliferation and activation of peritoneal NK cells, whose depletion by antibody reverses the antitumor effects [42]. Consistent with these findings, cathelicidin is expressed in tumor-infiltrating NK cells and is necessary for the tumor-suppressing activity of NK cells [50]. In mouse xenograft melanoma and lymphoma models, cathelicidin knockout mice (CRAMP$-/-$) showed faster tumor growth than wild type controls. NK cells derived from CRAMP$-/-$ mice demonstrated impaired cytotoxic activity against tumor cells [50].

In addition, LL-37 and human defensins play a pivotal role in the innate immune responses to *Helicobacter pylori* (Hp) infection. Increasing host defense against Hp prevents tumorigenesis of gastric carcinoma associated with Hp.

6   THE POTENTIAL FOR AMPs IN ANTICANCER THERAPY: OPPORTUNITIES AND CHALLENGES

AMPs can exert selective cytotoxic activities against cancer cells, essentially through their preference for partitioning into the plasma membrane of cancer cells. This stands in contrast to many types of anticancer therapies that target rapidly dividing cells. Even dormant cancer cells have negatively charged membranes; these cells could also be targeted by AMPs [58]. If, indeed, the positive charge on AMPs is the salient feature for their anticancer effects, it would be difficult for cancer cells develop chemoresistance against AMPs. For some types of cancer cells, disruption of the plasma membrane alone may be sufficient to account for the cytotoxicity of AMPs. In other types of cancer cells, AMPs must cross the plasma membrane and partition into the cytoplasm of cancer cells. This leads to inhibition of proliferation and induction of apoptosis through a variety of mechanisms including the disruption of the mitochondrial membrane, regulation of microRNA expression and gene transcription, and alteration of cancer cell metabolic profiles. In addition, some AMPs enhance host immune system-mediated anticancer activities. These anticancer properties of AMPs are not significantly affected by molecular heterogeneities within a given tumor or between different tumors, nor are they counteracted by the adaptability and alternative survival pathways in cancer cells, as frequently seen in cancer resistance. It may be that through their use of polypharmacology, AMPs are able to circumvent chemoresistance. Furthermore, as their membranolytic action is essentially independent of proliferation pathways, AMPs may exert cytotoxic activities against all cancer cells including nongrowing or slowly growing neoplastic cells. Lack of activity against these cell populations is part of the reason for development of chemoresistance and tumor recurrence.

Coadministration of AMPs with other types of chemotherapy may lead to enhanced cytotoxicity. The combination of LL-37 with TLR ligands such as CpG oligodeoxynucleotides shows synergistic toxicity against ovarian cancer cells [42]. AMPs can also be used to direct chemotherapeutic to mitochondria and enhance the cancer cytotoxicity (mitochondria targeted therapy). Cotreatment of HeLa cells with PEGylated liposomal formulation of epirubicin and the AMP hepcidin 2–3 significantly increased the cytotoxicity of epirubicin and reduced the expression of multidrug resistant (MDR) transporter-1 and -2 [59].

AMPs have potential as clinical diagnostic markers of GI cancer. This has been proposed with serum levels of HNP1–3 in colorectal cancer [16,17,19,21], tissue levels of HD6 in colon adenoma [18], serum levels of HD5 and HD6 in colon cancer [24,25], and tissue levels of HNP1–3 in gastric cancer [22].

To our knowledge, there has been only one clinical trial in the United States involving AMPs and cancer. "Induction of Antitumor Response in Melanoma Patients Using the Antimicrobial Peptide LL37" (NCT02225366) is an ongoing study scheduled to end in 2020. The researchers are examining the optimal biological dose of LL-37 based on both efficacy and toxicity. No results have been posted as of this writing.

However, there are significant challenges to overcome before AMPs can be used as an anticancer therapy in humans. First, as illustrated in Fig. 1, in vitro experiments reveal that LL-37 has both pro- and anticancer activities depending on the concentration used. The ideal therapy would be a carrier-assisted tumor-targeted system that could deliver effective anticancer concentrations of AMP to all tumor sites [60]. This may be one of the best options for using AMPs in cancer therapy. Nevertheless, more research needs to be done to understand the procancer mechanisms of low concentrations of AMPs.

A second challenge is that the actions of AMPs on tumors in vivo are even more complex than what is seen in vitro. The tumor environment may be infiltrated with AMP-secreting immune cells. Evidence suggests that the constitutively expressing AMPs and AMPs secreted from tumor infiltrating immune cells may play different roles. Constitutive expression of AMPs may prevent tumorigenesis, while once a tumor starts, tumor-derived AMPs may stimulate the recruitment of tumor stroma cells, which in turn produce more AMPs and other diverse factors, including inflammatory factors. This process promotes tumor growth and metastasis. In fact, the mixture of AMPs secreted by the immune cells may differ from those emanating from the cancer cells. The possibility of synergetic or antagonistic interactions between AMPs has not been explored. A more in-depth understanding of these intertwined pathways will be needed before clinical application of AMP anticancer therapy.

Third, because AMPs are rapidly metabolized and excreted, it will be necessary to find ways maintain therapeutic levels in vivo. This is complicated by other concerns as well. The pH of the tumor environment may favor AMP binding to the cancer cell plasma membrane and thus increase the potency of a particular AMP. Conversely, some AMPs may bind to anionic serum proteins within the tumor, therefore reducing the potency of these peptides. In addition, the introduction of exogenous AMPs into the tumor may also change the expression of the endogenous peptides which influence the effects of AMPs.

Finally, although polypharmacology might be considered an asset for AMPs when it comes to chemoresistance, the presence of multiple mechanisms of action could also be associated with some off-target toxicity. For example, LL-37 at high doses (25–50 μM) causes significant hemolysis [61,62] and is toxic to human leukocytes and the T-lymphocyte MOLT cell line [63]. The need to improve the selectivity of AMPs for cancer cells may require a better understanding of the structure-activity relationship of these molecules. Such an understanding would benefit the prospects of using assisted and targeted delivery of AMPs as a new therapeutic option for cancer treatment.

7   CONCLUSION

Do antimicrobial peptides represent a new era in chemotherapy for GI tumors? It is too soon to answer this question with any certainty. Certain features of AMPs make them attractive as novel anticancer agents. The chemical nature of these peptides—cationic and amphipathic—automatically directs them towards the important targets (active and dormant cancer cells) and limits their interaction with non-cancerous cells, regardless of their proliferation rate. The apparent ability of AMPs to attack cancer cells at several levels, polypharmacology, may act to thwart the evolution of chemoresistance within a tumor. Yet AMPs have significant problems. In particular, these peptides have both anti- and procancer properties that depend on both concentration and cell type. Even if we set aside the problems of how to administer and maintain the desired concentrations of AMPs at the tumor, we still do not have enough information to be certain of what the therapeutic dose should be and whether endogenous sources of AMPs and local cytokines might confound our efforts to control the dose. Although it sounds trivial to say "we need more research," we really do believe that "we need more research!" AMPs are selective for cancer cells but are promiscuous in their effects on cancer cell biology. Most individual studies have been focused on one particular mechanism of action; at some point it will be necessary to consider multiple mechanisms in the same system. The majority of the studies emphasize either cathelicidin or defensins, yet both are present in the human body as host defense molecules. Critical clues to the action of AMPs may come from comparisons of how they affect different types of cancer cells. Finally, it is encouraging to see that some research is geared towards modifying the amino acid sequence of AMPs with the goal of achieving more specific responses. We remain hopeful that AMPs will eventually become important contributors in the fight against cancer!

REFERENCES

[1] Niyonsaba F, Kiatsurayanon C, Chieosilapatham P, Ogawa H. Friends or foes? Host defense (antimicrobial) peptides and proteins in human skin diseases. Exp Dermatol 2017. https://doi.org/10.1111/exd.13314.
[2] Ren SX, Cheng AS, To KF, Tong JH, Li MS, Shen J, et al. Host immune defense peptide LL-37 activates caspase-independent apoptosis and suppresses colon cancer. Cancer Res 2012;72(24):6512–23. https://doi.org/10.1158/0008-5472.can-12-2359.
[3] Wu WK, Sung JJ, To KF, Yu L, Li HT, Li ZJ, et al. The host defense peptide LL-37 activates the tumor-suppressing bone morphogenetic protein signaling via inhibition of proteasome in gastric cancer cells. J Cell Physiol 2010;223(1):178–86. https://doi.org/10.1002/jcp.22026.
[4] Zaiou M, Gallo RL. Cathelicidins, essential gene-encoded mammalian antibiotics. J Mol Med 2002;80(9):549–61. https://doi.org/10.1007/s00109-002-0350-6.
[5] Zanetti M. The role of cathelicidins in the innate host defenses of mammals. Curr Issues Mol Biol 2005;7(2):179–96.
[6] Wu WK, Sung JJ, Cheng AS, Chan FK, Ng SS, To KF, et al. The Janus face of cathelicidin in tumorigenesis. Curr Med Chem 2014;21(21):2392–400.
[7] Gopal R, Jeong E, Seo CH, Park Y. Role of antimicrobial peptides expressed by host cells upon infection by Helicobacter pylori. Protein Pept Lett 2014;21(10):1057–64.
[8] Pero R, Coretti L, Nigro E, Lembo F, Laneri S, Lombardo B, et al. Beta-defensins in the fight against Helicobacter pylori. Molecules 2017;22(3). https://doi.org/10.3390/molecules22030424.
[9] Uraki S, Sugimoto K, Shiraki K, Tameda M, Inagaki Y, Ogura S, et al. Human beta-defensin-3 inhibits migration of colon cancer cells via downregulation of metastasis-associated 1 family, member 2 expression. Int J Oncol 2014;45(3):1059–64. https://doi.org/10.3892/ijo.2014.2507.
[10] Gudmundsson GH, Agerberth B, Odeberg J, Bergman T, Olsson B, Salcedo R. The human gene FALL39 and processing of the cathelin precursor to the antibacterial peptide LL-37 in granulocytes. Eur J Biochem 1996;238(2):325–32.
[11] Bandurska K, Berdowska A, Barczynska-Felusiak R, Krupa P. Unique features of human cathelicidin LL-37. Biofactors 2015;41(5):289–300. https://doi.org/10.1002/biof.1225.
[12] Hase K, Murakami M, Iimura M, Cole SP, Horibe Y, Ohtake T, et al. Expression of LL-37 by human gastric epithelial cells as a potential host defense mechanism against Helicobacter pylori. Gastroenterology 2003;125(6):1613–25.

[13] Li D, Liu W, Wang X, Wu J, Quan W, Yao Y, et al. Cathelicidin, an antimicrobial peptide produced by macrophages, promotes colon cancer by activating the Wnt/beta-catenin pathway. Oncotarget 2015;6(5):2939–50. https://doi.org/10.18632/oncotarget.2845.

[14] O'Neil DA, Porter EM, Elewaut D, Anderson GM, Eckmann L, Ganz T, Kagnoff MF. Expression and regulation of the human beta-defensins hBD-1 and hBD-2 in intestinal epithelium. J Immunol 1999;163(12):6718–24.

[15] Semlali A, Al Amri A, Azzi A, Al Shahrani O, Arafah M, Kohailan M, et al. Expression and new exon mutations of the human Beta defensins and their association on colon cancer development. PLoS One 2015;10(6). https://doi.org/10.1371/journal.pone.0126868.

[16] Albrethsen J, Bogebo R, Gammeltoft S, Olsen J, Winther B, Raskov H. Upregulated expression of human neutrophil peptides 1, 2 and 3 (HNP 1-3) in colon cancer serum and tumours: a biomarker study. BMC Cancer 2005;5. https://doi.org/10.1186/1471-2407-5-8.

[17] Albrethsen J, Moller CH, Olsen J, Raskov H, Gammeltoft S. Human neutrophil peptides 1, 2 and 3 are biochemical markers for metastatic colorectal cancer. Eur J Cancer 2006;42(17):3057–64. https://doi.org/10.1016/j.ejca.2006.05.039.

[18] Radeva MY, Jahns F, Wilhelm A, Glei M, Settmacher U, Greulich KO, Mothes H. Defensin alpha 6 (DEFA 6) overexpression threshold of over 60 fold can distinguish between adenoma and fully blown colon carcinoma in individual patients. BMC Cancer 2010;10. https://doi.org/10.1186/1471-2407-10-588.

[19] Melle C, Ernst G, Schimmel B, Bleul A, Thieme H, Kaufmann R, et al. Discovery and identification of alpha-defensins as low abundant, tumor-derived serum markers in colorectal cancer. Gastroenterology 2005;129(1):66–73.

[20] Pagnini C, Corleto VD, Mangoni ML, Pilozzi E, Torre MS, Marchese R, et al. Alteration of local microflora and alpha-defensins hyper-production in colonic adenoma mucosa. J Clin Gastroenterol 2011;45(7):602–10. https://doi.org/10.1097/MCG.0b013e31820abf29.

[21] Kemik O, Kemik AS, Sumer A, Begenik H, Purisa S, Tuzun S. Human neutrophil peptides 1, 2 and 3 (HNP 1-3): elevated serum levels in colorectal cancer and novel marker of lymphatic and hepatic metastasis. Hum Exp Toxicol 2013;32(2):167–71. https://doi.org/10.1177/0960327111412802.

[22] Cheng CC, Chang J, Chen LY, Ho AS, Huang KJ, Lee SC, et al. Human neutrophil peptides 1-3 as gastric cancer tissue markers measured by MALDI-imaging mass spectrometry: implications for infiltrated neutrophils as a tumor target. Dis Markers 2012;32(1):21–31. https://doi.org/10.3233/dma-2012-0857.

[23] Mohri Y, Mohri T, Wei W, Qi YJ, Martin A, Miki C, et al. Identification of macrophage migration inhibitory factor and human neutrophil peptides 1-3 as potential biomarkers for gastric cancer. Br J Cancer 2009;101(2):295–302. https://doi.org/10.1038/sj.bjc.6605138.

[24] Nam MJ, Kee MK, Kuick R, Hanash SM. Identification of defensin alpha6 as a potential biomarker in colon adenocarcinoma. J Biol Chem 2005;280(9):8260–5. https://doi.org/10.1074/jbc.M410054200.

[25] Nikitina IG, Bukurova Iu A, Hankin SL, Karpov VL, Lisitsyn NA, Beresten SF. Enteric alpha defensin 5 is secreted into the blood stream by colon tumors. Molekuliarnaia Biologiia (Mosk) 2013;47(1):133–6.

[26] Piktel E, Niemirowicz K, Wnorowska U, Watek M, Wollny T, Gluszek K, et al. The role of cathelicidin LL-37 in cancer development. Arch Immunol Ther Exp 2016;64(1):33–46. https://doi.org/10.1007/s00005-015-0359-5.

[27] Kuroda K, Fukuda T, Yoneyama H, Katayama M, Isogai H, Okumura K, Isogai E. Antiproliferative effect of an analogue of the LL-37 peptide in the colon cancer derived cell line HCT116 p53+/+ and p53. Oncol Rep 2012;28(3):829–34. https://doi.org/10.3892/or.2012.1876.

[28] Kuroda K, Fukuda T, Isogai H, Okumura K, Krstic-Demonacos M, Isogai E. Antimicrobial peptide FF/CAP18 induces apoptotic cell death in HCT116 colon cancer cells via changes in the metabolic profile. Int J Oncol 2015;46(4):1516–26. https://doi.org/10.3892/ijo.2015.2887.

[29] Kuroda K, Okumura K, Isogai H, Isogai E. The human cathelicidin antimicrobial peptide LL-37 and mimics are potential anticancer drugs. Front Oncol 2015;5. https://doi.org/10.3389/fonc.2015.00144.

[30] von Haussen J, Koczulla R, Shaykhiev R, Herr C, Pinkenburg O, Reimer D, et al. The host defence peptide LL-37/hCAP-18 is a growth factor for lung cancer cells. Lung Cancer 2008;59(1):12–23. https://doi.org/10.1016/j.lungcan.2007.07.014.

[31] Wang W, Jia J, Li C, Duan Q, Yang J, Wang X, et al. Antimicrobial peptide LL-37 promotes the proliferation and invasion of skin squamous cell carcinoma by upregulating DNA-binding protein A. Oncol Lett 2016;12(3):1745–52. https://doi.org/10.3892/ol.2016.4865.

[32] Jia J, Zheng Y, Wang W, Shao Y, Li Z, Wang Q, et al. Antimicrobial peptide LL-37 promotes YB-1 expression, and the viability, migration and invasion of malignant melanoma cells. Mol Med Rep 2017;15(1):240–8. https://doi.org/10.3892/mmr.2016.5978.

[33] Coffelt SB, Waterman RS, Florez L, Honer zu Bentrup K, Zwezdaryk KJ, Tomchuck SL, et al. Ovarian cancers overexpress the antimicrobial protein hCAP-18 and its derivative LL-37 increases ovarian cancer cell proliferation and invasion. Int J Cancer 2008;122(5):1030–9. https://doi.org/10.1002/ijc.23186.

[34] Girnita A, Zheng H, Gronberg A, Girnita L, Stahle M. Identification of the cathelicidin peptide LL-37 as agonist for the type I insulin-like growth factor receptor. Oncogene 2012;31(3):352–65. https://doi.org/10.1038/onc.2011.239.

[35] Weber G, Chamorro CI, Granath F, Liljegren A, Zreika S, Saidak Z, et al. Human antimicrobial protein hCAP18/LL-37 promotes a metastatic phenotype in breast cancer. Breast Cancer Res 2009;11(1):R6. https://doi.org/10.1186/bcr2221.

[36] Nishimura M, Abiko Y, Kurashige Y, Takeshima M, Yamazaki M, Kusano K, et al. Effect of defensin peptides on eukaryotic cells: primary epithelial cells, fibroblasts and squamous cell carcinoma cell lines. J Dermatol Sci 2004;36(2):87–95. https://doi.org/10.1016/j.jdermsci.2004.07.001.

[37] Shi N, Jin F, Zhang X, Clinton SK, Pan Z, Chen T. Overexpression of human beta-defensin 2 promotes growth and invasion during esophageal carcinogenesis. Oncotarget 2014;5(22):11333–44. https://doi.org/10.18632/oncotarget.2416.

[38] Muller CA, Markovic-Lipkovski J, Klatt T, Gamper J, Schwarz G, Beck H, et al. Human alpha-defensins HNPs-1, -2, and -3 in renal cell carcinoma: influences on tumor cell proliferation. Am J Pathol 2002;160(4):1311–24.

[39] Kuroda K, Fukuda T, Krstic-Demonacos M, Demonacos C, Okumura K, Isogai H, et al. miR-663a regulates growth of colon cancer cells, after administration of antimicrobial peptides, by targeting CXCR4-p21 pathway. BMC Cancer 2017;17(1):33. https://doi.org/10.1186/s12885-016-3003-9.

[40] Garcia-Quiroz J, Garcia-Becerra R, Santos-Martinez N, Avila E, Larrea F, Diaz L. Calcitriol stimulates gene expression of cathelicidin antimicrobial peptide in breast cancer cells with different phenotype. J Biomed Sci 2016;23(1):78. https://doi.org/10.1186/s12929-016-0298-4.

[41] Coffelt SB, Marini FC, Watson K, Zwezdaryk KJ, Dembinski JL, LaMarca HL, et al. The pro-inflammatory peptide LL-37 promotes ovarian tumor progression through recruitment of mul-tipotent mesenchymal stromal cells. Proc Natl Acad Sci U S A 2009;106(10):3806–11. https://doi.org/10.1073/pnas.0900244106.

[42] Chuang CM, Monie A, Wu A, Mao CP, Hung CF. Treatment with LL-37 peptide enhances antitumor effects induced by CpG oligodeoxynucleotides against ovarian cancer. Hum Gene Ther 2009;20(4):303–13. https://doi.org/10.1089/hum.2008.124.

[43] Sainz Jr B, Alcala S, Garcia E, Sanchez-Ripoll Y, Azevedo MM, Cioffi M, et al. Microenvironmental hCAP-18/LL-37 promotes pancreatic ductal adenocarcinoma by activating its cancer stem cell compartment. Gut 2015;64(12):1921–35. https://doi.org/10.1136/gutjnl-2014-308935.

[44] Hanaoka Y, Yamaguchi Y, Yamamoto H, Ishii M, Nagase T, Kurihara H, et al. In vitro and in vivo anticancer activity of human beta-defensin-3 and its mouse homolog. Anticancer Res 2016;36(11):5999–6004. https://doi.org/10.21873/anticanres.11188.

[45] Brogden KA. Antimicrobial peptides: pore formers or metabolic inhibitors in bacteria? Nat Rev Microbiol 2005;3(3):238–50. https://doi.org/10.1038/nrmicro1098.

[46] Riedl S, Zweytick D, Lohner K. Membrane-active host defense peptides—challenges and perspectives for the development of novel anticancer drugs. Chem Phys Lipids 2011;164(8):766–81. https://doi.org/10.1016/j.chemphyslip.2011.09.004.

[47] Lichtenstein A. Mechanism of mammalian cell lysis mediated by peptide defensins. Evidence for an initial alteration of the plasma membrane. J Clin Invest 1991;88(1):93–100. https://doi.org/10.1172/jci115310.

[48] Okumura K, Itoh A, Isogai E, Hirose K, Hosokawa Y, Abiko Y, et al. C-terminal domain of human CAP18 antimicrobial peptide induces apoptosis in oral squamous cell carcinoma SAS-H1 cells. Cancer Lett 2004;212(2):185–94. https://doi.org/10.1016/j.canlet.2004.04.006.

[49] Munoz M, Craske M, Severino P, de Lima TM, Labhart P, Chammas R, et al. Antimicrobial peptide LL-37 participates in the transcriptional regulation of melanoma cells. J Cancer 2016;7(15):2341–5. https://doi.org/10.7150/jca.16947.

[50] Buchau AS, Morizane S, Trowbridge J, Schauber J, Kotol P, Bui JD, Gallo RL. The host defense peptide cathelicidin is required for NK cell-mediated suppression of tumor growth. J Immunol 2010;184(1):369–78. https://doi.org/10.4049/jimmunol.0902110.

[51] Ren SX, Shen J, Cheng AS, Lu L, Chan RL, Li ZJ, et al. FK-16 derived from the anticancer peptide LL-37 induces caspase-independent apoptosis and autophagic cell death in colon cancer cells. PLoS One 2013;8(5). https://doi.org/10.1371/journal.pone.0063641.

[51a] Chi Y, Zhou D. MicroRNAs in colorectal carcinoma—from pathogenesis to therapy. J Exp Clin Cancer Res 2016;35:43–53.

[52] Bleuming SA, Kodach LL, Garcia Leon MJ, Richel DJ, Peppelenbosch MP, Reitsma PH, et al. Altered bone morphogenetic protein signalling in the Helicobacter pylori-infected stomach. J Pathol 2006;209(2):190–7. https://doi.org/10.1002/path.1976.

[53] Kodach LL, Wiercinska E, de Miranda NF, Bleuming SA, Musler AR, Peppelenbosch MP, et al. The bone morphogenetic protein pathway is inactivated in the majority of sporadic colorectal cancers. Gastroenterology 2008;134(5):1332–41. https://doi.org/10.1053/j.gastro.2008.02.059.

[54] Wen XZ, Miyake S, Akiyama Y, Yuasa Y. BMP-2 modulates the proliferation and differentiation of normal and cancerous gastric cells. Biochem Biophys Res Commun 2004;316(1):100–6. https://doi.org/10.1016/j.bbrc.2004.02.016.

[55] Haramis AP, Begthel H, van den Born M, van Es J, Jonkheer S, Offerhaus GJ, Clevers H. De novo crypt formation and juvenile polyposis on BMP inhibition in mouse intestine. Science 2004;303(5664):1684–6. https://doi.org/10.1126/science.1093587.

[56] Bleuming SA, He XC, Kodach LL, Hardwick JC, Koopman FA, Ten Kate FJ, et al. Bone morphogenetic protein signaling suppresses tumorigenesis at gastric epithelial transition zones in mice. Cancer Res 2007;67(17):8149–55. https://doi.org/10.1158/0008-5472.can-06-4659.

[57] Hurtado P, Peh CA. LL-37 promotes rapid sensing of CpG oligodeoxynucleotides by B lymphocytes and plasmacytoid dendritic cells. J Immunol 2010;184(3):1425–35. https://doi.org/10.4049/jimmunol.0902305.

[58] Baxter AA, Lay FT, Poon IKH, Kvansakul M, Hulett MD. Tumor cell membrane-targeting cationic antimicrobial peptides: novel insights into mechanisms of action and therapeutic prospects. Cell Mol Life Sci 2017. https://doi.org/10.1007/s00018-017-2604-z.

[59] Juang V, Lee HP, Lin AM, Lo YL. Cationic PEGylated liposomes incorporating an antimicrobial peptide tilapia hepcidin 2-3: an adjuvant of epirubicin to overcome multidrug resistance in cervical cancer cells. Int J Nanomedicine 2016;11:6047–64. https://doi.org/10.2147/ijn.s117618.

[60] Reinhardt A, Neundorf I. Design and application of antimicrobial peptide conjugates. Int J Mol Sci 2016;17(5). https://doi.org/10.3390/ijms17050701.

[61] Ciornei CD, Sigurdardottir T, Schmidtchen A, Bodelsson M. Antimicrobial and chemoattractant activity, lipopolysaccharide neutralization, cytotoxicity, and inhibition by serum of analogs of human cathelicidin LL-37. Antimicrob Agents Chemother 2005;49(7):2845–50. https://doi.org/10.1128/aac.49.7.2845-2850.2005.

[62] Oren Z, Lerman JC, Gudmundsson GH, Agerberth B, Shai Y. Structure and organization of the human antimicrobial peptide LL-37 in phospholipid membranes: relevance to the molecular basis for its non-cell-selective activity. Biochem J 1999;341(Pt 3):501–13.

[63] Johansson J, Gudmundsson GH, Rottenberg ME, Berndt KD, Agerberth B. Conformation-dependent antibacterial activity of the naturally occurring human peptide LL-37. J Biol Chem 1998;273(6):3718–24.

Chapter 6

AMPs and Mechanisms of Antimicrobial Action

Lucinda Furci and Massimiliano Secchi
San Raffaele Scientific Institute, Milan, Italy

Chapter Outline

1 INTRODUCTION

Multicellular organisms throughout evolution have developed immune-defense mechanisms that confer protection against invading microorganisms. Antimicrobial peptides (AMPs) are ribosome-synthesized peptides which rapidly eradicate or inactivate bacteria, viruses, fungi, and parasites. A diverse array of AMPs are made by nearly all multicellular organisms, including plants, insects, fish, and mammals. They also exert immunomodulatory and adjuvant functions by acting as chemotactic for immune cells, and inducing cytokines and chemokines secretion [1].

In mammals, the majority of AMPs are produced by the epithelial surfaces of tissues that interface directly with the environment, including skin,

Antimicrobial Peptides in Gastrointestinal Diseases. https://doi.org/10.1016/B978-0-12-814319-3.00006-4
Copyright © 2018 Chi Hin Cho. Published by Elsevier Ltd. All rights reserved.

respiratory tract, reproductive tract, and intestines. Each of these tissues frequently encounters a diverse array of microorganisms that can cause disease. In addition, such bodily sites are associated with a resident microbiota composed largely of commensal microorganisms. Thus, surface tissues are faced with the enormous challenge of maintaining homeostasis with complex assemblies of commensal bacteria while also limiting pathogen invasion [2]. In general, the intestinal mucosa plays a critical role in host protection by a repertoire of antimicrobial proteins that provide protection from food and water-borne pathogens and help to shape the composition of gut microbiota [2].

2 COMMON STRUCTURAL FEATURES OF AMPs

AMPs are primarily cationic, amphipathic peptides with broad-spectrum microbicidal activity mainly associated with membrane permeabilization. They are small molecules between 30 and 150 amino acids, with an hydrophobic content of around 60% and the net charge varies between +2 and +11 [3], suggesting a highly conserved structure.

Secondary structures of human AMPs may include α-helices (like LL-37), β-sheets (like α-defensin 1), and extended peptides (like indolicidin) [4]. AMPs with linear α-helices are notably positively charged and amphipathic, but only adopt such secondary conformation upon binding to bacterial membranes [5]. Therefore, their antimicrobial activity is directly dependent on the secondary structure. In contrast, β-sheet AMPs possess well-defined conformations due to the presence of disulfide bridges formed between thiol groups in cysteines. In these peptides, such bridges are often required for antimicrobial activity [6]. Other intestinal AMPs, such as human-defensins 3, present both α-helices and β-sheets [7].

As another common feature, most intestinal AMPs (α-defensins, REG3α and LL37) are produced as inactive propeptides and some are even stored inside epithelial cell secretory granules. This is to fine tune AMP's activity and to suppress membrane-toxicity during storage inside the cells [8]. Propiece moieties are very sensitive to proteases and are generally removed by proteolytic activity once the peptide is released from the cell.

3 OVERVIEW AND LOCALIZATION

Virtually all intestinal antimicrobial peptides derive from four distinct sources: (i) Paneth cells. Paneth cells express constitutively a large array of AMPs, of which α-defensins are the most abundant; (ii) Intestinal epithelial cells. The intestinal epithelium produces β-defensins and other important lectins upon stimulation during infections; (iii) Infiltrating neutrophils. That release an abundant stockpile of α-defensins 1–4 (HNP 1–4) at the site of infection; and (iv) Commensal bacteria. The mass of commensal bacteria inhabiting in the mucosa produce a vast array of bacteriocins (see Fig. 1).

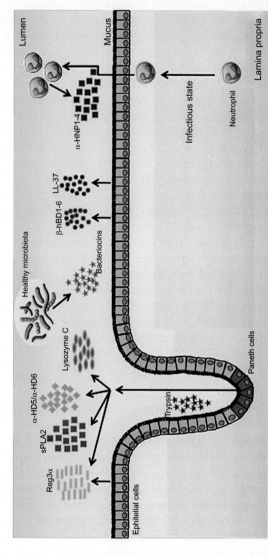

FIG.1 Sources of antimicrobial peptides (AMPs) in the intestinal mucosa. Virtually all the intestinal antimicrobial peptides derive from four distinct sources: a) Paneth cells express constitutively and store in their granules, lysozyme, enteric α-defensins (pro-HD5 and pro-HD6), and REG3α together with the serine protease trypsin. All these molecules are synthesized as inactive propeptides that will get activated by proteolytic cleavage by trypsin, in the crypt's lumen; b) The intestinal epithelium produces β-defensins 1-6, LL37, REGα and other important lectins (see later) in response to pathogens; c) infiltrating neutrophils release an abundant stockpile of HNPs; d) lastly, the mass of commensal bacteria inhabiting the intestinal mucosa produce a vast array of bacteriocins.

3.1   Paneth Cells as AMP Factory

Paneth cells are epithelial cells with intensive secretory activity located in small clusters at the base of the crypts of Lieberkuhn. Their large secretory granules contain massive concentrations of AMPs [2,9], proteases, and a labile zinc pool of unknown function [10,11]. Paneth cells produce AMPs constitutively and store them as inactive peptides in their secretory granules for release at later time [12,13]. Degranulation has been reported in response to bacteria and bacterial antigens and appears to happen in a dose-dependent fashion [14]. Several studies suggested that Paneth cells directly sense commensal bacteria of the gut through the MyD88/toll-like receptor (TLR) pathway [15]. Indeed, various components of the bacterial cell wall have been shown to trigger Paneth cell degranulation. Among these are lipopolysaccharides, lipotechoic acid, lipid A, and muramyl dipeptide [14], all of which stimulate TLRs. AMPs release from Paneth is also triggered by proinflammatory signals, as in the case of IFN-γ [16], or directly by cholinergic agonists [17]. Thus, we can say that Paneth's AMPs are constitutively expressed but their release by degranulation cells is highly regulated.

Paneth cells discharge their granules into the crypts lumen, where propeptides are activated by proteolytic cleavage (mainly by trypsin) [18] and diffuse into the mucus layer and lumen. This distribution provides antimicrobial protection to the crypts and contributes to the mucosal barrier from the small intestine to the colon [19]. Indeed, Paneth cells are located only in the duodenum, jejunum, and ileum, but Paneth's peptides can travel embedded in the upper layer of intestinal mucus [20] up to the colon. Paneth antimicrobial peptides, include three main categories: (i) enteric α-defensins (HD5 and HD6) [21,22]; (ii) enzymes (lysozyme C [23] and phospholipases [24]); (iii) lectins, like the C-type lectin REG3α [25].

3.1.1   Enteric α-Defensins

Defensins are the major family of AMPs of mammals [26–29]. While diverse in overall sequence and structure, they are related through six invariant cysteine residues that form 3 intramolecular disulfide bridges found in all defensins [30]. The presence of three disulfide bonds in such small peptides make these antimicrobials extremely stable and resistant to protelytic activity. Indeed, α-defensins, in their immature pro-form, are highly sensitive to proteases, that activates them by cleavage of the propiece; however, in their mature-form, defensins become highly resistant to any proteolysis. Additionally, there is a highly conserved salt bridge formed between Arg6-Glu14 that is thought to provide further stability against proteolytic activity but that is not required for the antimicrobial activity [31].

Based on sequence homology and disulfide topology, human defensins are classified into α and β families. Myeloid α-defensins 1–4, also called Human Neutrophil Peptides 1–4 (HNP1–4), are expressed predominantly by neutrophils, commonly making up as much as 50% of the total protein content in these cells [32,33]. Enteric α-defensins 5 and 6 (HD5 and HD6) are expressed copackaged in the granules of intestinal Paneth cells [34,35]. HD5 is the most abundant Paneth cells AMP [36]. Indeed, the transcriptional levels of HD5 are reported to be four- to sixfold greater than that of HD6, and higher than any other antimicrobial peptide in Paneth cell granules [37]. Although their expression follows the distribution of Paneth cells in duodenum, jejunum, and ileum; the highest expression levels are observed in the ileum [35].

Human enteric defensin-5 (HD5) is a 32-residue peptide with an overall charge of +4 at neutral pH. HD5 is expressed as a 94-residue prepropeptide that consists of an N-terminal 19-residue signal sequence that directs the peptide to the secretory pathway, a 43-residue proregion, and a 32-residue C-terminal region that corresponds to the mature peptide [35]. Biophysical studies on the quaternary structure revealed that HD5 forms dimers at pH 4 and tetramers at neutral pH [38]. HD5 is recognized as an AMP that can kill a broad range of microbes [39–43]. Seminal investigations that employed a HD5 transgenic mouse revealed that HD5 expression in the small intestine confers resistance to oral *Salmonella enterica* serovar Typhimurium challenge [44]. Subsequent studies of these mice supported a homeostatic role of HD5 in modulating the composition of the intestinal microbiota [45]. Taken together, these observations indicate that HD5 is important for both defense against pathogens and maintaining intestinal homeostasis.

Human enteric defensin-6 (HD6), like HD5, is expressed and stored in the Paneth cells granules as inactive 81-aa propeptide. HD6 is also cleaved by trypsin and results in a mature 32-aa peptide with an overall charge +2 at neutral pH [46] and a tertiary structure similar to HD5 [34,40]. Surprisingly, while mature HD5 has potent bactericidal activity *in vitro*, the second Paneth α-defensin HD6, is devoid of bactericidal activity [40,47,48], at least while its tri-disulfide bonds are oxidized [49]. Indeed, when the crystal structure of HD6 was initially solved, the intermolecular interactions were found to be different from other alpha-defensins [48]. The significance of this difference became clear in 2012, when a new and unprecedented model for HD6 was presented: rather than killing microbes, HD6 self-assembles into higher-order oligomers that are named "nanonets" to entrap pathogens and prevent them from invading host cells [47].

3.1.2 Lysozyme

Lysozyme is an enzyme that specifically hydrolyzes peptidoglycans. This enzyme is very abundant in the surface lining fluid of many organs, including the intestine where it is chiefly made by Paneth cells. In addition, lysozyme is abundant in macrophages and neutrophils. Its high concentration in these cells and at mucosal surfaces supports that lysozyme has a substantial role as an antimicrobial agent *in vivo*, even though specific activity *in vitro* is modest [50].

3.1.3 Phosholipase A2

Phospholipase A2 enzymes are a large family of catalytic molecules that hydrolyze the fatty acid ester bond at position sn-2 of membrane phosphotriglycerides. Paneth cells granules contain abundant quantities of one specific member of this family that has selective activity on bacterial membranes [51–53]. Colonic epithelial cells also express this enzyme, which is named secretory phospholipase A2 (sPLA2) [54]. Like lysozyme, sPLA2 is not only expressed in epithelial cells, but also in macrophages.

3.1.4 REG3α

REG3α is a soluble C-type lectin with a single carbohydrate recognition domain expressed in the human intestine [55]. This lectin is abundantly expressed in Paneth cells and enterocytes [36,56,57]. The expression of REG3α while is constitutive in Paneth cells can be induced by injury and inflammatory signals in the epithelium [58].

3.1.5 Intelectin-1

Another lectin abundantly expressed in Paneth cells is intelectin-1 (IntL-1) [62,63], which is encoded by INTL1, a gene that was identified in genetic screens for inflammatory bowel disease (IBD)-susceptibility loci [64]. Intelectins are classified as X-type lectins, that exhibit a calcium-dependent function. IntLs contain a fibrinogen-like domain (FBD) and are proposed to function similarly to ficolins, a class of FBD containing lectins that contribute to innate immunity [65]. In mammals, IntLs are predominantly produced by lung cells, intestinal goblet cells, and intestinal Paneth cells during infection [66,67].

3.2 AMPs From Enterocytes and Colonocytes

In addition to Paneth that release the larger part of intestinal antimicrobial peptides, other AMPs deriving from the mucosal epithelium, both from enterocytes and colonocytes, contribute to the protection of the intestinal milieu against invading microorganisms.

3.2.1 Human β-Defensins

While α-defensins production is restricted to the crypts of Lieberkühn, β-defensins are expressed throughout the epithelium in both the small and large intestines [68]. Eight of the 28 β-defensins encoded in the human genome are expressed in intestinal epithelial cells [69,70]. These defensins are longer than α-defensins, spanning from 36–50 amino acids, but also share three disulfide bounds and a triple-stranded β-sheet conformation. Of note, unlike α-defensins human β-defensins do not form pores and higher order oligomers in solution [28]. Also, contrary to α-defensins that are constitutively expressed, intestinal epithelial expression of β-defensins is inducible, except for the apparent constitutive expression of HBD-1. Therefore, whenever there is an increase in the demand of AMP activity (e.g., upon infection or tissue damage) their concentrations must surge to reach the required protective levels. Of note, an excessive increase in the concentration of any AMP may lead to tissue damage and local functions may be compromised.

3.2.2 Cathelicidins (LL37)

Cathelicidins are a very diverse group of cationic α-helical and amphipathic AMPs that exhibit broad-spectrum activity against bacteria, fungi, and viruses [1]. There is only one human gene encoding multiple cathelicidin-related peptides [71,72] located on chromosome 3 and expressed in the airways, mouth, tongue, esophagus, and small intestine [73,74]. LL37 is highly positively charged (+6 charge at physiological pH 7.4), due to the high content of arginine and lysine amino acids, and adopts an α-helical structure in solutions with ionic composition similar to that of human plasma [75]. Cathelicidins are characterized by a highly conserved N-terminal signal peptide (the so-called cathelin domain) and a highly variable C-terminal antimicrobial domain that can be released after cleavage by proteases [72,76].

3.2.3 Lipocalin-2

Lipocalin-2 is an AMP that binds bacterial siderophores, which are the catechol-related iron chelators secreted by bacteria to acquire iron from their environment. Sequestering these iron chelators prevents this mode of bacterial iron acquisition and inhibits bacterial growth [77]. Many cells express lipocalin-2 inducibly, including enterocytes and macrophages [78–80]. In the intestine, lipocalin-2 may selectively inhibit growth of siderocalin dependent bacteria at the mucosal surface.

3.3 AMPs From Immune Cells

Besides the production of AMPs by Paneth cells and enterocytes, it is crucial to consider that immune cells normally patrolling the intestinal surface can significantly contribute to the homeostasis of AMPs. Neutrophils are the major producers of myeloid α-defensins HNP 1–4, which are stored as precursors proteins at very high concentration in their azurophilic granules, commonly making up as much as 50% of their total protein content [32,33]. HNP1–4 are not expressed in the gut, however activated neutrophils have been shown to reach the intestinal lumen by transmigration across the intestinal epithelium and release a large amount of mature HNPs. Indeed, during acute inflammatory processes, excessive polymorphonuclear leukocytes transepithelial migration and a massive release of AMPs can damage mucosal integrity and lead to the development of pathological intestinal inflammation [81–86].

3.4 AMPs From Microbiota

Bacteriocines: It should be mentioned that in the intestinal microenvironment, the host is not the sole source of AMPs. The intestinal microbiota has been identified as a rich source of potential probiotic bacteria that produce antimicrobial bacteriocins that exert a beneficial effect on the gastrointestinal tract (GI) while specifically inhibiting GI pathogens [87]. Bacteriocins are bacterial ribosomally synthesized AMPs which are often modified posttranslationally, like most host antimicrobial peptides [88,89].

Different bacteria produce different types of bacteriocins that potentially reach high concentrations in certain local regions of the gut. These compounds act in a nontargeted manner and their contribution to probiotic functionality has not been investigated as extensively. For instance, *Lactobacillus* spp. in the human intestine, produce antimicrobial factors, including lantibiotics, small heat-stable, nonlanthionine containing membrane-active peptides, larger heat-labile proteins, and complex bacteriocins containing one or more chemical moieties [90,91]. Probiotics are producing diverse antimicrobial agents and may be beneficial for the treatment and prevention of a variety of infectious diseases caused by oral, enteric pathogens and urogenital infections [92]. These strains may represent a valid alternative therapeutic alternatives in the multidrug-resistant pathogens [93].

4 MECHANISMS OF MICROBIAL KILLING

As a matter of fact, the vast majority of intestinal AMPs, targets bacterial membranes. The evolutionary rationale for targeting essential cell-wall structures lies in the fact that such structures are difficult for microorganisms to modify without a consequent loss in overall fitness. This reduces the likelihood that bacteria will develop resistance to such AMPs.

4.1 Bacterial Membrane Permeabilization

In terms of structure, AMPs represent an extremely diverse group of biological active molecules but all shear cationic and amphipathic properties [6] that make them suitable for membrane targeting. The basic and hydrophobic amino acids are about 50% more abundant in AMPs than in genomes overall, while acidic and polar amino acids are about 75% less abundant than expected. In line with these observations, recent literature suggests that antimicrobial activity is not dependent on specific amino acid sequences or on specific peptide structures [94–96]. Instead, activity depends more on the amino acid composition of the peptide and on its physical chemical properties, see Table 1.

The mixed cationic and hydrophobic composition of AMPs makes them well-suited for interacting with and perturbing microbial cytoplasmic membranes that typically present anionic surfaces, rich in lipids, such as

TABLE 1 Main Intestinal Antimicrobial Peptides and Their Mechanism of Action

Producing Cells	Antimicrobial Peptide	Mechanism of Action
Paneth cells	HD5	Membrane damage, Lectin activity
	HD6	Self-assembles "nanonets"
	sPLA2	Enzymatic activity
	Lysozyme C	Enzymatic activity
	Reg3α	Lectin activity
	Intelectin-1	Lectin activity
Intestinal epithelium	hBD1-6	Membrane disruption
	LL-37	Membrane disruption
	Lipocalin 2	Binding of bacterial siderophores
	Intelectin-1	Lectin activity
	Reg3α	Lectin activity
Immune cells		
Neutrophils	HNP1–4	Membrane damage, inhibition cell wall synthesis
Macrophages	Lysozyme C	Enzymatic activity
Microbiota		
Gram-negative	Colicines	Different killing mechanism
	Microcines	Different killing mechanism
Gram-positive	Class I	Prevent cell wall synthesis
	Class II	Membrane permeabilization
	Class III	Different killing mechanism
	Class IV	Different killing mechanism

phosphatidyl glycerol or cardiolipin, to the outside environment. The fact that all Gram-negative and Gram-positive bacteria display these type of negatively charged lipids accounts for the lack of specificity of most AMPs and promotes the attraction between AMPs and bacterial membranes while preventing their binding to most host cells membranes. One of the most commonly cited explanations for the selectivity of AMPs for microbes over host cells is the difference in membrane interactions due to differences in exposed anionic lipid content. Binding of AMPs to microbial membranes is significant, while binding of AMPs to the neutral phosphatidylcholine/cholesterol/sphingomyelin-rich surfaces of animal plasma membranes is weaker. Another hypothesis to explain the selective toxicity of AMPs relies on differences in the membrane potential of microbes and mammalian cells. Microbes tend to have a significantly large charge difference across their membranes compared to mammalian cells, which favor cationic defensins to selectively target microbes [82,97–99].

4.1.1   Timing in Bacterial Membrane Permeabilization

There is ample direct evidence that most AMPs permeabilize microbial cytoplasmic membranes and that the membranes are often permeabilized with increasing severity with time [100]. AMPs can dissipate the electrochemical gradient across microbial plasma membranes within a few seconds of addition [100,101]. This implies that AMPs are able to rapidly pass through the thick proteoglycan layer of Gram-positive bacteria or the outer membrane lipopolysaccharide layer of Gram-negative bacteria in few seconds. Permeation of larger markers, including dye markers, metabolites, and cytosolic proteins, through the cytoplasmic membrane occurs on the time scale of minutes to tens of minutes [94,101]. After an hour or more in contact with AMPs, gross disruption of microbial membrane structure and morphology is often noted, including membrane blebbing, vesiculation, fragmentation, release of DNA, cell aggregation, and destruction of cell morphology.

4.1.2   Specific Steps in AMPs-Membrane Interaction

Regardless of the time required, or the specific antimicrobial mechanism, defined steps must occur to induce bacterial killing.

4.1.2.1   Attraction

Antimicrobial peptides must first be attracted to bacterial surfaces, and one obvious mechanism is electrostatic bonding between anionic or cationic peptides and structures on the bacterial surface. Cationic AMPs are likely to first be attracted to the net negative charges that exist on the outer envelope of Gram-negative bacteria—for example, anionic phospholipids and phosphate groups on LPS—and to the teichoic acids on the surface of Gram-positive bacteria.

4.1.2.2   Attachment

Once close to the microbial surface, peptides have to pass through capsular polysaccharides before they can interact with the outer membrane, which contains LPS in Gram-negative bacteria, and traverse capsular polysaccharides, teichoic acids, and lipoteichoic acids before they can interact with the cytoplasmic membrane in Gram-positive bacteria. Then they have gained access to the cytoplasmic membrane and they can interact with lipid bilayers. *In vitro* studies of antimicrobial peptides incubated with artificial membranes or vesicles show that peptides bind in two physically distinct states [6,102]. At low peptide/lipid ratios, α-helical peptides and β-sheet peptides, adsorb and embed into the lipid head group region in a functionally inactive state that stretches the membrane [103]. The extent of membrane thinning is specific to the peptide and directly proportional to the peptide concentration. An example of this phenomenon is mesosome-like structures that form in the plasma membrane of *C. difficile* bacilli treated with HD5 [42]. Mesosomes are produced by the lateral expansion of the membrane area occurring upon binding and insertion of the AMPs [104]. Such spherical or tubular double-layered membrane structures extend inward from the cell surface and are visualized by electron microscopy [104] (see Fig. 2).

4.1.2.3 Peptide-Insertion

The interaction AMPs/membrane is dependent from the peptide concentration. At low peptide-to-lipid ratio, AMPs tend to embed into membrane or adsorb onto the surface, adopting an orientation parallel to the membrane bilayer [105]. Increasing the peptide-to-lipid ratio, AMPs adopt a perpendicular orientation relative to the bacterial membrane, which allows them to insert themselves into the membrane and promote clustering. Different models of interaction have been proposed: (i) Barrel-stave pore model, in which the AMPs form dimers or multimers that cross the membrane forming barrel-like channels [106]; (ii) Toroidal pore model, in which the peptide forms a monolayer by connecting the outer and the inner lipid layers in the pore [107]; and (iii) Carpet model, where AMPs form a carpet-like structure covering the outer surface of the membrane acting like detergents disrupting the bacterial membrane [108].

However, although descriptions of membrane damage seem to vary, they are likely to be related. Ion channels, transmembrane pores, and extensive membrane rupture do not represent three completely different modes of action, but instead are a continuous graduation between them [109]; beginning at low peptide concentration, with transmembrane potential, pH gradient are destroyed and respiration is inhibited, to complete membrane disruption [110–112].

4.2 Enzymatic Attack on Bacterial Wall Structures

Several epithelial AMPs kill bacteria through enzymatic attack on key structures of the cell wall. In the GI tract two of these AMP-enzymes are lysozyme and sPLA2.

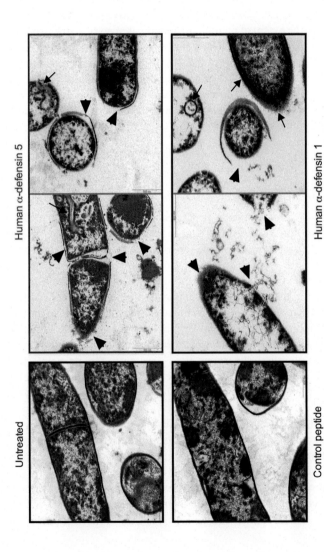

FIG. 2 Bactericidal effects of HD5 and HNP1 on *C. difficile*. Transmission electron micrographs of α-defensin-treated *C. difficile*. Suspensions of *C. difficile* CD630 bacilli incubated in the absence or in the presence of 7 μM HD5, or HNP1, or control peptide RL26495; a concentrations that can be normally found in the small intestine [15]. Arrowheads indicate cell wall detachment or severe leakage of cytoplasmic contents; arrows indicate double-layered mesosome-like structures and fibers extending from the cell surface.

4.2.1 Lysozyme

Lysozyme is a glycosidase that hydrolyzes the 1,4-*b*-glycosidic linkages between the *N*-acetylglucosamine and *N*-acetylmuramic acid moieties of peptidoglycan. Consequently, lysozyme is more effective against Gram-positive bacteria, where the peptidoglycan is more accessible, than against Gram-negative organisms, where the peptidoglycan is protected by the outer membrane [113]. In addition to directly killing bacteria, lysozyme enzymatic activity can regulate downstream innate immune responses to certain microorganisms. Knockout mice points to another key *in vivo* role for lysozyme—an ability to impede peptidoglycan accumulation in tissues. By enzymatic degradation of peptidoglycan, lysozyme activity appears to help avoid prolonged inflammatory responses that otherwise would result from persistence of bacterial cell wall antigens [50]. Thus, whether eliminating peptidoglycan from crypts and intestinal tissue, or antibacterial activity against lumenal microbes, is the chief role for intestinal lysozyme remains an open question.

4.2.2 Secretory Phospholipase A2 (sPLA2)

Secretory phospholipase A2 (sPLA2) is a second example of an epithelial AMP that kills bacteria through an enzymatic mechanism. Bacterial membranes, rich in phosphatidylglycerol and phosphatidylethanolamine, are the key targets of sPLA2, but the enzyme can cleave other phosphotriglyceride substrates [52,114]. sPLA2 penetrates the bacterial cell wall to gain access to the membrane, where it hydrolyzes phospholipids and thus compromises bacterial membrane integrity [115]. This enzyme is bactericidal, with preferential activity against Gram positive bacteria [114,116].

4.3 Lectins as Bacterial Killers

Within the array of intestinal AMPs few lectins are present. Lectins are carbohydrate-binding proteins that contribute to mucosal innate immunity.

4.3.1 REG3α

REG3α binds to the peptidoglycan of the bacterial cell wall [117,118]. REG3α bactericidal activity is therefore selective for Gram-positive bacteria because peptidoglycan is generally accessible on the outer surfaces of Gram-positive bacteria but is shielded by the outer membrane in Gram-negative bacteria [56,61]. REG3α recognition of peptidoglycan involves binding to the extended carbohydrate chains on the bacterial surfaces, but not to shorter, soluble peptidoglycan chains that are shed by bacteria. This selectivity for extended carbohydrate chains allows REG3α to interact with the bacterial surface and not be sequestered by shed glycans [117].

After binding to the peptidoglycan, REG3α permeabilizes the bacterial membrane by utilizing its cationic residues to interact with the negatively charged bacterial membrane. REG3α, upon contact with lipids, oligomerizes to form hexameric transmembrane pores. Recently as been shown that REG3α, at high concentrations, can form higher-order macromolecular assemblies of hexameric pores that result in inactive filamentous structures. The formation of higher-order filamentous assemblies is frequently observed for mammalian membrane toxic proteins in other tissue contexts, such as the brain and the nervous system [59,60]. This suggests that filamentation might sequester the membrane-active REG3α pores and limit the toxicity toward the intestinal epithelium once a microbial threat has been eliminated [61].

4.3.2 Human Intelectin-1

Human intelectin-1 (hIntL-1) is a secretory glycoprotein consisting of polypeptides and N-linked oligosaccharides. It binds galactofuranose, a sugar found only in microorganisms, and hence may serve as a microbial pattern element [66]. Recent studies demonstrated that hIntL-1 selectively binds glycans with terminal 1,2-diol groups [63], providing new molecular-level insight into how it recognizes binding partners. In total, hIntL-1 exemplifies how lectins can distinguish microbes from mammalian cells in the innate immune response.

4.4 Interference at Intracellular Level

While most AMPs interact with and influence the integrity of microbial membranes, it is not known if membrane permeabilization is always the lethal event or if the membrane is the only site of action. It seems likely that many AMPs can translocate across microbial membranes at concentrations that do not induce permeabilization. Once in the cytoplasm, they can target DNA and chaperonins, alter the cytoplasmic membrane septum formation, inhibit cell-wall synthesis, reduce nucleic-acid synthesis, suppress protein synthesis, or inhibit enzymatic activity [119–121]. Multiple simultaneous mechanisms of AMP action (e.g., membrane permeabilization as well as intracellular effects) may help explain their broad-spectrum activity and the rarity of inducible resistance. For instance, human β-defensin 2 (HBD2) has been recently found to localize at septal foci of *Enterococcus faecalis* and disrupt virulence factor assembly [122].

4.4.1 Bacterial Cell Filamentation

Bacterial cell filamentation is a phenomenon that has been reported for neutrophil peptides HNP1–2 and enteric defensin HD5. Cells exposed to these peptides have an extremely elongated morphology, which indicates that the peptide-treated cells are unable to undergo cell division. It is not known whether cell filamentation is due to the blocking of DNA replication or the inhibition of membrane proteins that are involved in septum formation [123,124].

4.4.2 Inhibition DNA, RNA, and Protein Synthesis

Among AMPs, α-helical peptides, proline- and arginine-rich peptides, and defensins have all been shown to be able to block (3H)thymidine, (3H)uridine, and (3H)leucine uptake in *E. coli*; thus indirectly showing that they inhibit DNA, RNA, and protein synthesis [125–127]. Wherease, a direct effect on DNA, RNA, and protein synthesis, as been shown for HNP-1 and -2 [125].

4.4.3 Inhibition of Cell Wall Synthesis

The bactericidal mechanism of HNP1 [128] and β-defensin 3 (HBD3) [129] has been found to involve to varying degrees [130] the interaction with the peptidoglycan precursor lipid II and inhibition of cell wall synthesis [128].

4.5 Bacterial Trapping by Nanonets

HD6 lacks direct bactericidal activity when its six cysteines are oxidized to form three intramolecular disulfide bonds [49,131]. Recent studies have identified a nonbactericidal role for HD6 in host defense. HD6 spontaneously self-assembles into multipeptide nanonets upon contact with bacterial appendages such as flagella and fimbrae [47]. This function gives HD6 a key role in aggregating and sequestering bacteria that enter the crypts of the small intestine; rather than direct cell killing [47,49]. Proof of principle of this mechanism has been obtained in transgenic mice expressing HD6 that exhibit resistance to oral *S. typhimurium* challenge. In this model the *Salmonella* burden in the feces remains high [47], indicating that the observed protective effect does not result from antibacterial activity against this pathogen. In further support to the notion that HD6 entraps bacterial pathogens, HD6 blocks the ability of *S. typhimurium* [47] and *Listeria monocytogenes* [132] to invade cultured epithelial cells. Moreover, the N-terminal proregion of proHD6 blocks self-assembly and functional activity [34,133]. Interestingly, HD6 and HD5 although cosecreted by Paneth cells, do not synergize in terms of antibacterial activity rather, HD6 specifically and synergistically enhances the HD5-induced IL-8 secretion [134].

4.6 AMPs Exploiting Multiple Antimicrobial Strategies

Up to now, a complex behavior in terms of mechanism of action has been demonstrated only for a few AMPs. It is very likely that this is the case for a large number of AMPs. Further studies are needed to reveal such complexity in order to gain enough knowledge to be transferred to the development of new therapeutic agents.

4.6.1 α-Defensin 5

Results from several recent studies indicate that HD5 function is multifaceted and that the peptide has the capacity to affect microbes in ways other than cell killing [135].

Membrane destabilization: Unlikely most AMPs, HD5 shows a different effect with wall structures of different bacteria. For instance *C. difficile* treated with HD5 undergoes wall and membrane rupture and pronounced cell fragmentation, as documented by FACS analysis and electron microscopy; while, *E. coli*, in the same study, displays only increase in cell volume and membrane depolarization [42]. Accordingly, Chileveru et al. reports that *E. coli* treated with HD5 exhibits cell elongation and cellular blebs but not membrane rupture. Similar morphological changes occur for other Gram-negative bacteria, including *Acinetobacter baumannii* and *Pseudomonas aeruginosa* [124]. From these studies, authors inferred that during the attack of HD5 on Gram-negative bacteria, HD5 crosses the outer membrane, permeabilizes the inner membrane, and kills the bacteria via a nonlytic mechanism of action [124,131].

Bacterial cell filamentation: The extremely elongated morphology of *E. coli* cells exposed to HD5 indicates that the peptide-treated cells are unable to undergo cell division and has been interpreted as cell filamentation. This phenomenon is not only due to peptide insertion into the membrane but also to the blocking of DNA replication and the inhibition of membrane proteins involved in septum formation [6,42,136].

Lectin-like behavior and antitoxin activity: Several evidences address the lectin-like behavior of HD5. HD5 binds membrane glycoproteins, causes clumping of bacterial cells [124], and inhibit HIV-1 infection by binding to the carbohydrate moieties of viral glycoproteins, such as gp120 to the host receptor CD4 on T cells [43,137].

Most importantly, the lectin behavior allows HD5 to bind and inactivate several bacterial toxins such as anthrax lethal factor and *C. difficile* enterotoxin B [138,139]. New evidence suggests that reduced HD5 may act to sequester and neutralize free bacterial LPS in the gut, blunting the associated inflammatory response [140]. Moreover, a recent study showed that interaction of HD5 with a *Vibrio colera* toxin induced toxin unfolding and enhanced the susceptibility of the toxin to proteolysis. These data suggest a new model of interaction of HD5 with bacterial toxins whereby unfoldase activity results in stabilization of non-native states of bacterial toxins, resulting in aggregation or proteolysis [141].

4.6.2 Cathelicidins (LL37)

LL-37 was shown to disrupt bacterial membranes through the formation of toroidal pores and carpet structures [142], and to exhibit chemotactic properties, attracting leukocytes and activating secretion of chemokines. In addition, LL-37 is internalized by cells, acidified in endosomes, and activates the signaling pathway downstream to Toll-like receptor 3 by interacting with double-stranded RNA [143].

4.6.3 Bacteriocines

Bacteriocins share most common mechanisms of bacterial killing, like destruction of target cells by pore formation and/or inhibition of cell wall synthesis, with host AMPs [144]. Bacteriocins are active against numerous foodborne and human pathogens, are produced by "generally regarded as safe" microorganisms—like Lactobacilli—and are readily degraded by proteolytic host systems, which make them attractive candidates for biotechnological applications [88,145,146].

5 INFLUENCE OF THE PHYSIOLOGICAL MILIEU

Effective definitions of AMP activity and specificity should take into consideration the physiological conditions *in vivo*. This includes the concentrations of AMPs at the sites of infection, the role of synergistic substances that might be present in tissues and fluids (e.g., the presence of lysozyme, other AMPs and proteins, and the absence of divalent cations), the role of inhibiting substances that might be present (e.g., physiological concentrations of salts and serous proteins) and the unusual characteristics of bacteria replicating *in vivo*, particularly those in biofilms [147]. For instance, the *in vitro* antimicrobial activity of many defensins is attenuated by the presence of salt and divalent cations, like millimolar concentrations of sodium chloride [39,148]. This phenomenon is commonly attributed to a disruption of electrostatic interactions between the cationic defensin and anionic bacterial cell membrane [6]. Divalent cations like Ca(II), Mg(II), and Zn(II), also block defensin activity. This observation is of broad interest because the Paneth cell granules, which express and store HD5, also contain a labile zinc pool of unknown function [10].

An important challenge to overcome is the lower AMP's activity *in vivo* when compared to that observed *in vitro*. This has largely hampered the development of AMPs as therapeutic agents, as this decrease in activity stems from differences in pH and salt concentrations, the presence of proteases and corresponding inhibitors, and other interacting molecules hindering antimicrobial activity [149]. Despite very effective *in vitro*, both hBD-1 and hBD-2 present attenuated activities *in vivo* as a result of their sensitivity to physiological salinity. In contrast, hBD-3 is active against *S. aureus* and vancomycin-resistant *E. faecium* even at physiological salt concentrations [150].

An important influencing factor is pH. In the presence of a reducing agent, HD6 becomes more hydrophobic and this reduced form inhibits the growth of *Bifidobacterium adolescentis* [49]. This suggests that local and momentary conditions within the physiological environment may be able to direct the action of defensins in a dynamic fashion. Recently, it was also found that the effectiveness of HD1–4 as antimicrobials varies with the pH and reducing conditions of the environment [49]. This suggests that defensins have certain conditions in which they function optimally, as such, environmental factors may play a role in the regulation of the activity of enteric defensins.

6 ANTIMICROBIAL PEPTIDES AS POTENTIAL THERAPEUTIC AGENTS

6.1 HD5 vs *Salmonella typhimurium*

Proof of principle for HD5-associated antimicrobial activity *in vivo* has been obtained with transgenic mice by showing a direct cause-and-effect relation between the presence or absence of HD5 expression and survival of infection with the enteric pathogen *Salmonella enterica* serovar Typhimurium [44]. Matrix metalloproteinase-7 (MMP-7), like trypsin in humans, processes mouse pro-α-defensins in its active forms and is therefore essential for α-defensin antimicrobial activity. Mmp7-knockout mice exhibit enhanced susceptibility to an oral challenge with *Salmonella typhimurium*, suggesting that fully functional α-defensin-5 is required for mucosal protection against intestinal bacterial pathogens [151]. Supporting this idea, transgenic mice that overexpress human α-defensin-5 in Paneth cells show greater resistance to oral *S. typhimurium* challenge than wild type mice [44].

6.2 HD5 and HNP-1 vs *Clostridium difficile*

The disease: Clostridium difficile, a Gram-positive, spore-forming anaerobic bacterium, is considered the major known cause of health care-associated infectious diarrhea in Western countries [152]. The disease spectrum caused by *C. difficile* infection (CDI) ranges from mild diarrhea to severe pseudomembranous enterocolitis, sepsis, and death [153]. *C. difficile* is transmitted via endospores that resist the acidity of the stomach and germinate in the small intestine, resulting vegetative cells colonizing in the colon. They can reside there asymptomatically for a long time [154].

The pathogen: Disruption of the normal gut microflora by broad-spectrum antibiotics [155] triggers *C. difficile* proliferation and causes disease through the production of cytotoxic toxins A and B [156]. It has been observed the epidemic diffusion in hospitals and health care settings of *C. difficile* strains referred to be hypervirulent, that is, ribotypes 027, 078, and 018 [157]. Each of them carries one or more virulence factors, such as production of binary toxin, mutation in regulatory toxin genes tcdC and tcdD, or fluoroquinolone (FQ) resistance, and all are strongly associated with increased severity of CDI and higher attributable mortality [158–160]. *C. difficile* is also resistant to several AMPs. The susceptibility to defensins was not an obvious finding: cationic AMPs share a mechanism of action, and *C. difficile* has evolved numerous strategies to evade their attack [161]. *C. difficile* has been described to be resistant to bacterially derived AMPs, like bacitracin, nisin, gallidermin, vancomycin, and polymyxin B, but also to host-derived AMPs, like lysozyme [162–164]. Furthermore, resistance to mammalian SMAP-29 and LL-37 was reported for epidemic-associated PCR ribotype 027 isolates [165].

The intestinal environment: HD5 in the intestine is the most abundant AMP: it has been estimated that up to $450\,\mu g/cm^2$ is stored in the ileal mucosa, with concentrations of 14–$70\,\mu M$ [18]. Intestinal microbiota homeostasis is maintained by the dynamic interplay between AMPs, mainly HD5, and commensal bacteria [45]. HD5 controls the enteric microbiota composition by selective killing of bacterial pathogens while preserving commensals [2]; in turn, resident bacteria stimulate HD5 production via Toll-like receptor (TLR)-MyD88 signaling [15]. In the course of CDI, the interaction of *C. difficile* toxins with colonic cells triggers a significant inflammatory response and neutrophil accumulation at the site of epithelial damage, with a massive release of HNP1 [161]. Several studies have demonstrated that neutrophils are critical for defense against *C. difficile* infection [166,167]. The observation that *C. difficile* can reside asymptomatically in the intestines of immunocompetent individuals, whereas severe *C. difficile* associated disease (CDAD) occurs mainly in immunocompromised or elderly subjects [153], strongly suggests that host immune responses are important determinants of disease pathogenesis [156]. HD5 not only restricts *C. difficile* colonization by maintaining microbiota homeostasis and inactivating *C. difficile* toxin B [139], but also as described here, directly kills *C. difficile* bacilli.

The findings: In this study it was shown for the first time that human α-defensins exert potent dose-dependent damage to the vegetative isoform of *C. difficile*, resulting in plasma membrane depolarization and bacterial fragmentation. This damage was documented by electron microscopy (see Fig. 2) and quantified by FACS analysis. The size of the bacilli and their degree of membrane depolarization (measured as uptake of Oxonol) could be simultaneously visualized by FACS analysis. Treatment with both HD5 and HPN-1 resulted in a log increase of oxonol fluorescence, indicating massive membrane depolarization and the appearance of a new population in small cellular debris. All strains tested were highly susceptible to the microbicidal activity of both α-defensins, with epidemic hypervirulent isolates being among the most sensitive to the microbicidal activity of α-defensins [42].

The results show that *C. difficile* is susceptible to both HD5 and HNP1, with IC_{50} in the nanomolar range; thus HD5, which, in the small intestine, can reach concentrations of $70\,\mu M$ [18], is likely to largely block the replication of clostridia at the site of germination [154], suggesting a protective role against *C. difficile* colonization. Most importantly, all strains belonging to PCR ribotype 027 were shown to be twice as susceptible to both HNP1 and HD5 as reference strain CD630 and almost 10 times more susceptible to HD5 than was *E. faecalis* [42]. These evidences clearly suggest that α-defensins can circumvent the mechanisms of evasion adopted by *C. difficile* to resist catheli-cidin LL-37 [165]. A plausible explanation relies on cationic peptide structure: LL-37 is characterized by an extended α-helical structure and can be cleaved and inactivated by bacterial proteases [115], while mature HD5 and HNP1, due to their tightly folded structure, are inherently resistant to proteolysis [20,168].

These findings also address the notion of a peculiar mechanism of interaction between α-defensins and *C. difficile* bacilli [42]. Indeed, transmission electron microscopy analysis showed morphological alterations of the bacterial cell wall and cytoplasmic membrane consistent with the cationic, amphiphilic nature of α-defensins, which are electrostatically attracted by the negatively charged bacterial surface layers and get embedded into the hydrophobic regions of the lipid membranes [96,169] (see Fig. 2).

However, we did not observe blisters, protruding bubbles, and overall moderate damage to the bacterial wall as reported for other Gram-positive bacteria [104,170,171]; rather, we found massive damage of the bacterial cell wall and plasma membrane, with leakage of cytoplasmic content and widespread cell fragmentation [42]. Accordingly, data from FACS analysis showed a pattern of damage peculiar to *C. difficile*: bacterial fragmentation was completely absent in *E. coli*, consistent with its Gram-negative nature [124,171], but was also absent in *E. faecalis* [42], a Gram-positive organism also characterized by a thick peptidoglycan wall.

HD5 was more potent than HNP1 at a lower range of concentrations, that is, 0.3 and 3 μM, whereas maximal concentrations of these two defensins resulted in similar bactericidal activities, with slightly more strength for HNP1 [42]. This agrees with the propensity of HD5 to form aggregates at high concentrations, thus losing available sites to interact with the cell membrane [172].

Furthermore, such a different range of activities is compatible with the different physiological roles of HD5 and HNP1. Indeed, HD5, which is secreted at high concentrations in the intestinal crypts, gets diluted in the mucous layer, and still maintains its bactericidal activity [14,20]. On the other hand, HNP1, whose role is to intervene once the inflammatory process is initiated and a massive bacterial invasion has to be tackled, is more active at the highest levels of the range (7 μM). Accordingly, physiological concentrations of HNP1 are very high, both in neutrophils (above 10 mg/ml) and in neutrophils nets, where it kills, respectively, engulfed and trapped bacteria [173], thus indicating that levels of HNP1 microbicidal for *C. difficile* bacilli can be easily reached in the extracellular milieu in the vicinity of activated neutrophils (Fig. 3).

Hence, we can speculate that in immunocompetent individuals, *C. difficile* spores germinate in the small intestine, where bacilli encounter HD5 at concentrations more than 50 times higher than their $IC_{50}s$ [15]. Then the surviving bacilli travel to the colon carried by the mucous flow, rich with HD5 and other AMPs [19], which constrains their replicative capacity and protects the host from *C. difficile* colonization (Fig. 3A). This hypothesis is supported by experiments showing that HD5 persists in an intact and functional form throughout

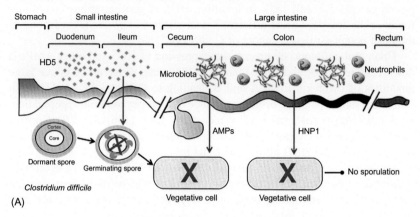

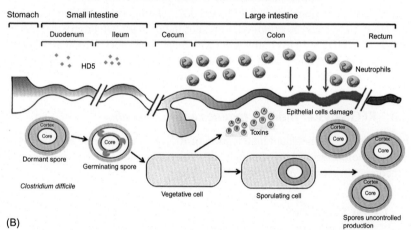

FIG.3 How AMPs control *C. difficile* colonization of the gut. (A) Immunocompetent host. Dormant *C. difficile* spores, once ingested travel to the duodenum were start to germinate upon contact with bile salts. Here bacilli encounter HD5 at concentrations 50 times higher than their IC_{50}. Surviving bacilli travel to the colon carried by the mucous flow, rich in HD5, other AMPs, HNP1–4 from some neutrophil and bacteriocins from microbiota, which constrains their replicative capacity and protects the host from *C. difficile* colonization. (B) Immunocompromized host. Upon oral antibiotic treatment, the microbiota is reduced to a minimum and the expression of HD5 and other AMPs is downregulated [175] but *C.difficile*, that is resistant to most of the commonly used antibiotics, starts to proliferate without constraints. Colonizing bacilli produce enterotoxins A and B that cause great damage to the intestinal epithelium resulting in *C. difficile* infection (CDI). In response to this insult neutrophils infiltrate massively into the mucosa resulting in further damage and *C. difficile* associated disease (CDAD).

the all intestinal tract, including the colon [20], and is active throughout a broad pH range (pH 5.5–8.0) [39]. In addition, HD5 at concentrations commonly found in the small intestine efficiently neutralizes *C. difficile* toxin B, one of the most potent virulence factors of *C. difficile* [139].

Upon oral antibiotic treatment, the expression of HD5 and other AMPs is downregulated [174] and pathogenic *C. difficile* can thrive and produce large quantities of cytotoxic toxin A and B, leading to CDI and CDAD [154]. In immunodeficient patients, or in the absence of a proper bacterial repopulation, HD5 deficiency persists and leads to recurrent CDAD [156] (Fig. 3B). Concurrently, during early stages of CDI, neutrophil infiltration and release of HNP1 at the site of infection [167] play a beneficial role for the clearance of *C. difficile* bacilli, whereas in the case of advanced stages of CDAD, massive neutrophil infiltration enhances the inflammatory response and leads to host damage [161]. Notably, neutrophils, unlike macrophages and lymphocytes, are also resistant to *C. difficile* toxin A-mediated apoptosis [156].

From this point of view, the fact that *C. difficile*, especially highly virulent epidemic strains like ribotypes 018, 078, and 027, is highly susceptible to both HD5 and HNP1 [42], could be exploited to prevent and/or treat CDI. From a therapeutic perspective, HD5 used in combination with fecal microbiota transplant therapies [175] would contribute, with its antitoxin, bactericidal, and immunostimulatory actions, to the treatment of detrimental forms of recurrent CDAD [176].

6.3 REG3γ vs *L. monocytogenes* and *E. faecalis*

Finally, antibody-mediated inactivation of REG3γ reveals a role for REG3γ in limiting colonization by intestinal pathogens such as *L. monocytogenes* and vancomycin-resistant *E. faecalis* [177,178].

7 CONCLUSIONS

Bacterial infections and the considerable rise in antibiotic resistance in hospital and community settings pose significant problems for global health initiatives [179,180]. Infectious diseases caused by bacteria, viruses, or fungi are among the leading causes of death worldwide. The emergence of drug-resistance mechanisms, especially among bacteria, threatens the efficacy of all current antimicrobial agents, some of them already ineffective. As a result, there is an urgent need for new antimicrobial drugs.

The mammalian intestinal epithelium is faced with a continuous and complex microbial challenge that is unique among tissues. Epithelial cells cope with this challenge in part by producing a diverse array of AMPs that rapidly kill or inactivate bacteria. The emerging picture is that epithelial AMPs not only protect against pathogen colonization and invasion but also shape the composition and physical location of indigenous bacterial communities. Fully illuminating how AMPs impact the microbial ecology of the intestine will require further work.

Finally, the majority of *in vivo* studies performed to date have focused on intestinal AMP activities against bacteria. However, the intestinal microbiota also includes eukaryotic viruses [181], bacteriophages [182], and eukaryotic organisms such as fungi [183]. It will be interesting to determine how intestinal AMPs influence these other elements of the microbial community, either directly by targeting of these microbes or indirectly by affecting the bacterial communities in the intestine.

REFERENCES

[1] Zasloff M. Antimicrobial peptides of multicellular organisms. Nature 2002;415(6870):389–95. Epub 2002/01/25. https://doi.org/10.1038/415389a PubMed PMID: 11807545.

[2] Bevins CL, Salzman NH. Paneth cells, antimicrobial peptides and maintenance of intestinal homeostasis. Nat Rev Microbiol 2011;9(5):356–68. Epub 2011/03/23. https://doi.org/10.1038/nrmicro2546. PubMed PMID: 21423246.

[3] Wang G. Human antimicrobial peptides and proteins. Pharmaceuticals (Basel) 2014;7 (5):545–94. https://doi.org/10.3390/ph7050545. PubMed PMID: 24828484; PubMed Central PMCID: PMCPMC4035769.

[4] Lee DG, Kim HK, Kim SA, Park Y, Park SC, Jang SH, et al. Fungicidal effect of indolicidin and its interaction with phospholipid membranes. Biochem Biophys Res Commun 2003;305 (2):305–10. PubMed PMID: 12745074.

[5] Tossi A, Sandri L, Giangaspero A. Amphipathic, alpha-helical antimicrobial peptides. Biopolymers 2000;55(1):4–30. https://doi.org/10.1002/1097-0282(2000)55:1<4::AID-BIP30>3.0.CO;2-M. PubMed PMID: 10931439.

[6] Brogden KA. Antimicrobial peptides: pore formers or metabolic inhibitors in bacteria? Nat Rev 2005;3(3):238–50. https://doi.org/10.1038/nrmicro1098. PubMed PMID: 15703760.

[7] Kluver E, Schulz-Maronde S, Scheid S, Meyer B, Forssmann WG, Adermann K. Structure-activity relation of human beta-defensin 3: influence of disulfide bonds and cysteine substitution on antimicrobial activity and cytotoxicity. Biochemistry 2005;44(28):9804–16. https://doi.org/10.1021/bi050272k. PubMed PMID: 16008365.

[8] Lichtenstein A, Ganz T, Selsted ME, Lehrer RI. In vitro tumor cell cytolysis mediated by peptide defensins of human and rabbit granulocytes. Blood 1986;68(6):1407–10. PubMed PMID: 3779104.

[9] Clevers HC, Bevins CL. Paneth cells: maestros of the small intestinal crypts. Annu Rev Physiol 2013;75:289–311. https://doi.org/10.1146/annurev-physiol-030212-183744. PubMed PMID: 23398152.

[10] Giblin LJ, Chang CJ, Bentley AF, Frederickson C, Lippard SJ, Frederickson CJ. Zinc-secreting Paneth cells studied by ZP fluorescence. J Histochem Cytochem 2006;54 (3):311–6. https://doi.org/10.1369/jhc.5A6724.2005. PubMed PMID: 16260591.

[11] Dinsdale D. Ultrastructural localization of zinc and calcium within the granules of rat Paneth cells. J Histochem Cytochem 1984;32(2):139–45. PubMed PMID: 6693753.

[12] Cunliffe RN, Rose FR, Keyte J, Abberley L, Chan WC, Mahida YR. Human defensin 5 is stored in precursor form in normal Paneth cells and is expressed by some villous epithelial cells and by metaplastic Paneth cells in the colon in inflammatory bowel disease. Gut 2001;48(2):176–85.

[13] Ouellette AJ. Paneth cell alpha-defensins in enteric innate immunity. Cell Mol Life Sci 2011;68(13):2215–29. https://doi.org/10.1007/s00018-011-0714-6. PubMed PMID: 21560070.

[14] Ayabe T, Satchell DP, Wilson CL, Parks WC, Selsted ME, Ouellette AJ. Secretion of micro-bicidal alpha-defensins by intestinal Paneth cells in response to bacteria. Nat Immunol 2000; 1(2):113–8. PubMed PMID: 11248802.

[15] Vaishnava S, Behrendt CL, Ismail AS, Eckmann L, Hooper LV. Paneth cells directly sense gut commensals and maintain homeostasis at the intestinal host-microbial interface. Proc Natl Acad Sci U S A 2008;105(52):20858–63. Epub 2008/12/17. https://doi.org/10.1073/pnas. 0808723105 PubMed PMID: 19075245; PubMed Central PMCID: PMC2603261.

[16] Farin HF, Karthaus WR, Kujala P, Rakhshandehroo M, Schwank G, Vries RG, et al. Paneth cell extrusion and release of antimicrobial products is directly controlled by immune cell-derived IFN-gamma. J Exp Med 2014;211(7):1393–405. https://doi.org/10.1084/jem.20130753. PubMed PMID: 24980747PubMed Central PMCID: PMCPMC4076587.

[17] Satoh Y, Habara Y, Ono K, Kanno T. Carbamylcholine- and catecholamine-induced intracel-lular calcium dynamics of epithelial cells in mouse ileal crypts. Gastroenterology 1995; 108(5):1345–56. PubMed PMID: 7729625.

[18] Ghosh D, Porter E, Shen B, Lee SK, Wilk D, Drazba J, et al. Paneth cell trypsin is the proces-sing enzyme for human defensin-5. Nat Immunol 2002;3(6):583–90.

[19] Meyer-Hoffert U, Hornef MW, Henriques-Normark B, Axelsson LG, Midtvedt T, Putsep K, et al. Secreted enteric antimicrobial activity localises to the mucus surface layer. Gut 2008;57 (6):764–71. Epub 2008/02/06. https://doi.org/10.1136/gut.2007.141481. PubMed PMID: 18250125.

[20] Mastroianni JR, Ouellette AJ. Alpha-defensins in enteric innate immunity: functional Paneth cell alpha-defensins in mouse colonic lumen. J Biol Chem 2009;284(41):27848–56. Epub 2009/08/19. https://doi.org/10.1074/jbc.M109.050773. PubMed PMID: 19687006; PubMed Central PMCID: PMC2788835.

[21] Ouellette AJ, Darmoul D, Tran D, Huttner KM, Yuan J, Selsted ME. Peptide localization and gene structure of cryptdin 4, a differentially expressed mouse paneth cell alpha-defensin. Infect Immun 1999;67(12):6643–51. PubMed PMID: 10569786; PubMed Central PMCID: PMCPMC97078.

[22] Porter EM, Liu L, Oren A, Anton PA, Ganz T. Localization of human intestinal defensin 5 in Paneth cell granules. Infect Immun 1997;65(6):2389–95.

[23] Ghoos Y, Vantrappen G. The cytochemical localization of lysozyme in Paneth cell granules. Histochem J 1971;3(3):175–8. PubMed PMID: 4106573.

[24] Kiyohara H, Egami H, Shibata Y, Murata K, Ohshima S, Ogawa M. Light microscopic immu-nohistochemical analysis of the distribution of group II phospholipase A2 in human digestive organs. J Histochem Cytochem 1992;40(11):1659–64. https://doi.org/10.1177/40.11.1431054. PubMed PMID: 1431054.

[25] Lasserre C, Colnot C, Brechot C, Poirier F. HIP/PAP gene, encoding a C-type lectin over-expressed in primary liver cancer, is expressed in nervous system as well as in intestine and pancreas of the postimplantation mouse embryo. Am J Pathol 1999;154(5):1601–10. https://doi.org/10.1016/S0002-9440(10)65413-2. PubMed PMID: 10329612; PubMed Cen-tral PMCID: PMCPMC1866603.

[26] Ganz T. Defensins: antimicrobial peptides of innate immunity. Nat Rev Immunol 2003; 3(9):710–20. PubMed PMID: 12949495.

[27] Lehrer RI. Primate defensins. Nature reviews 2004;2(9):727–38.

[28] Selsted ME, Ouellette AJ. Mammalian defensins in the antimicrobial immune response. Nat Immunol 2005;6(6):551–7.

[29] Lehrer RI, Lu W. alpha-Defensins in human innate immunity. Immunol Rev 2012;245 (1):84–112. Epub 2011/12/16. https://doi.org/10.1111/j.1600-065X.2011.01082.x. PubMed PMID: 22168415.

[30] Xie C, Prahl A, Ericksen B, Wu Z, Zeng P, Li X, et al. Reconstruction of the conserved beta-bulge in mammalian defensins using D-amino acids. J Biol Chem 2005;280(38):32921–9. https://doi.org/10.1074/jbc.M503084200. PubMed PMID: 15894545.

[31] Rajabi M, de Leeuw E, Pazgier M, Li J, Lubkowski J, Lu W. The conserved salt bridge in human alpha-defensin 5 is required for its precursor processing and proteolytic stability. J Biol Chem 2008. PubMed PMID: 18499668.

[32] Date Y, Nakazato M, Shiomi K, Toshimori H, Kangawa K, Matsuo H, et al. Localization of human neutrophil peptide (HNP) and its messenger RNA in neutrophil series. Ann Hematol 1994;69(2):73–7. PubMed PMID: 8080882.

[33] Faurschou M, Borregaard N. Neutrophil granules and secretory vesicles in inflammation. Microbes Infect 2003;5(14):1317–27. PubMed PMID: 14613775.

[34] Chairatana P, Chu H, Castillo PA, Shen B, Bevins CL, Nolan EM. Proteolysis triggers self-assembly and unmasks innate immune function of a human alpha-defensin peptide. Chem Sci 2016;7(3):1738–52. https://doi.org/10.1039/C5SC04194E. PubMed PMID: 27076903; PubMed Central PMCID: PMCPMC4827351.

[35] Jones DE, Bevins CL. Paneth cells of the human small intestine express an antimicrobial peptide gene. J Biol Chem 1992;267(32):23216–25. PubMed PMID: 1429669.

[36] Wehkamp J, Salzman NH, Porter E, Nuding S, Weichenthal M, Petras RE, et al. Reduced Paneth cell alpha-defensins in ileal Crohn's disease. Proc Natl Acad Sci U S A 2005; 102(50):18129–34. PubMed PMID: 16330776.

[37] Wehkamp J, Chu H, Shen B, Feathers RW, Kays RJ, Lee SK, et al. Paneth cell antimicrobial peptides: topographical distribution and quantification in human gastrointestinal tissues. FEBS Lett 2006;580(22):5344–50. PubMed PMID: 16989824.

[38] Wommack AJ, Robson SA, Wanniarachchi YA, Wan A, Turner CJ, Wagner G, et al. NMR solution structure and condition-dependent oligomerization of the antimicrobial peptide human defensin 5. Biochemistry 2012;51(48):9624–37. https://doi.org/10.1021/bi301255u. PubMed PMID: 23163963; PubMed Central PMCID: PMCPMC3579768.

[39] Porter EM, van Dam E, Valore EV, Ganz T. Broad-spectrum antimicrobial activity of human intestinal defensin 5. Infect Immun 1997;65(6):2396–401.

[40] Ericksen B, Wu Z, Lu W, Lehrer RI. Antibacterial activity and specificity of the six human {alpha}-defensins. Antimicrob Agents Chemother 2005;49(1):269–75. PubMed PMID: 15616305.

[41] Nuding S, Zabel LT, Enders C, Porter E, Fellermann K, Wehkamp J, et al. Antibacterial activity of human defensins on anaerobic intestinal bacterial species: a major role of HBD-3. Microbes Infect 2009;11(3):384–93. Epub 2009/04/29. https://doi.org/10.1016/j.micinf. 2009.01.001 PubMed PMID: 19397883.

[42] Furci L, Baldan R, Bianchini V, Trovato A, Ossi C, Cichero P, et al. New role for human alpha-defensin 5 in the fight against hypervirulent clostridium difficile strains. Infect Immun 2015;83(3):986–95. https://doi.org/10.1128/IAI.02955-14. PubMed PMID: 25547793.

[43] Furci L, Tolazzi M, Sironi F, Vassena L, Lusso P. Inhibition of HIV-1 infection by human alpha-defensin-5, a natural antimicrobial peptide expressed in the genital and intestinal muco-sae. PLoS One 2012;7(9):e45208. Epub 2012/10/03. https://doi.org/10.1371/journal.pone. 0045208. PubMed PMID: 23028850; PubMed Central PMCID: PMC3459904.

[44] Salzman NH, Ghosh D, Huttner KM, Paterson Y, Bevins CL. Protection against enteric sal-monellosis in transgenic mice expressing a human intestinal defensin. Nature 2003;422 (6931):522–6.

[45] Salzman NH, Hung K, Haribhai D, Chu H, Karlsson-Sjoberg J, Amir E, et al. Enteric defen-sins are essential regulators of intestinal microbial ecology. Nat Immunol 2010;11(1):76–83.

Epub 2009/10/27. https://doi.org/10.1038/ni.1825. PubMed PMID: 19855381; PubMed Central PMCID: PMC2795796.

[46] Jones DE, Bevins CL. Defensin-6 mRNA in human Paneth cells: implications for antimicrobial peptides in host defense of the human bowel. FEBS Lett 1993;315(2):187–92. PubMed PMID: 8417977.

[47] Chu H, Pazgier M, Jung G, Nuccio SP, Castillo PA, de Jong MF, et al. Human alpha-defensin 6 promotes mucosal innate immunity through self-assembled peptide nanonets. Science (New York, NY) 2012;337(6093):477–81. https://doi.org/10.1126/science.1218831. PubMed PMID: 22722251.

[48] Szyk A, Wu Z, Tucker K, Yang D, Lu W, Lubkowski J. Crystal structures of human alpha-defensins HNP4, HD5, and HD6. Protein Sci 2006;15(12):2749–60. PubMed PMID: 17088326.

[49] Schroeder BO, Ehmann D, Precht JC, Castillo PA, Kuchler R, Berger J, et al. Paneth cell alpha-defensin 6 (HD-6) is an antimicrobial peptide. Mucosal Immunol 2015;8(3):661–71. https://doi.org/10.1038/mi.2014.100. PubMed PMID: 25354318; PubMed Central PMCID: PMCPMC4424388.

[50] Ganz T, Gabayan V, Liao HI, Liu L, Oren A, Graf T, et al. Increased inflammation in lysozyme M-deficient mice in response to Micrococcus luteus and its peptidoglycan. Blood 2003; 101(6):2388–92. https://doi.org/10.1182/blood-2002-07-2319. PubMed PMID: 12411294.

[51] Lambeau G, Gelb MH. Biochemistry and physiology of mammalian secreted phospholipases A2. Annu Rev Biochem 2008;77:495–520. https://doi.org/10.1146/annurev.biochem.76.062405.154007. PubMed PMID: 18405237.

[52] Murakami M, Taketomi Y, Girard C, Yamamoto K, Lambeau G. Emerging roles of secreted phospholipase A2 enzymes: lessons from transgenic and knockout mice. Biochimie 2010; 92(6):561–82. https://doi.org/10.1016/j.biochi.2010.03.015. PubMed PMID: 20347923.

[53] Vadas P, Pruzanski W. Induction of group II phospholipase A2 expression and pathogenesis of the sepsis syndrome. Circ Shock 1993;39(2):160–7. PubMed PMID: 8490995.

[54] Qu XD, Lloyd KC, Walsh JH, Lehrer RI. Secretion of type II phospholipase A2 and cryptdin by rat small intestinal Paneth cells. Infect Immun 1996;64(12):5161–5. PubMed PMID: 8945560; PubMed Central PMCID: PMCPMC174502.

[55] Choi SM, McAleer JP, Zheng M, Pociask DA, Kaplan MH, Qin S, et al. Innate Stat3-mediated induction of the antimicrobial protein Reg3gamma is required for host defense against MRSA pneumonia. J Exp Med 2013;210(3):551–61. https://doi.org/10.1084/jem.20120260. PubMed PMID: 23401489.

[56] Cash HL, Whitham CV, Behrendt CL, Hooper LV. Symbiotic bacteria direct expression of an intestinal bactericidal lectin. Science (New York, NY) 2006;313(5790):1126–30. Epub 2006/08/26. https://doi.org/10.1126/science.1127119 PubMed PMID: 16931762; PubMed Central PMCID: PMC2716667.

[57] Christa L, Carnot F, Simon MT, Levavasseur F, Stinnakre MG, Lasserre C, et al. HIP/PAP is an adhesive protein expressed in hepatocarcinoma, normal Paneth, and pancreatic cells. Am J Physiol 1996;271(6 Pt 1):G993–1002. https://doi.org/10.1152/ajpgi.1996.271.6.G993. PubMed PMID: 8997243.

[58] Lai Y, Li D, Li C, Muehleisen B, Radek KA, Park HJ, et al. The antimicrobial protein REG3A regulates keratinocyte proliferation and differentiation after skin injury. Immunity 2012; 37(1):74–84. https://doi.org/10.1016/j.immuni.2012.04.010. PubMed PMID: 22727489; PubMed Central PMCID: PMCPMC3828049.

[59] Butterfield SM, Lashuel HA. Amyloidogenic protein-membrane interactions: mechanistic insight from model systems. Angew Chem Int Ed Engl 2010;49(33):5628–54. https://doi.org/10.1002/anie.200906670. PubMed PMID: 20623810.

[60] Kagan BL, Jang H, Capone R, Teran Arce F, Ramachandran S, Lal R, et al. Antimicrobial properties of amyloid peptides. Mol Pharm 2012;9(4):708–17. https://doi.org/10.1021/mp200419b. PubMed PMID: 22081976; PubMed Central PMCID: PMCPMC3297685.

[61] Mukherjee S, Zheng H, Derebe MG, Callenberg KM, Partch CL, Rollins D, et al. Antibacterial membrane attack by a pore-forming intestinal C-type lectin. Nature 2014;505 (7481):103–7. https://doi.org/10.1038/nature12729. PubMed PMID: 24256734; PubMed Central PMCID: PMCPMC4160023.

[62] Komiya T, Tanigawa Y, Hirohashi S. Cloning of the novel gene intelectin, which is expressed in intestinal paneth cells in mice. Biochem Biophys Res Commun 1998;251(3):759–62. https://doi.org/10.1006/bbrc.1998.9513. PubMed PMID: 9790983.

[63] Wesener DA, Wangkanont K, McBride R, Song X, Kraft MB, Hodges HL, et al. Recognition of microbial glycans by human intelectin-1. Nat Struct Mol Biol 2015;22(8):603–10. https://doi.org/10.1038/nsmb.3053. PubMed PMID: 26148048; PubMed Central PMCID: PMCPMC4526365.

[64] Barrett JC, Hansoul S, Nicolae DL, Cho JH, Duerr RH, Rioux JD, et al. Genome-wide association defines more than 30 distinct susceptibility loci for Crohn's disease. Nat Genet 2008; 40(8):955–62. https://doi.org/10.1038/ng.175. PubMed PMID: 18587394; PubMed Central PMCID: PMCPMC2574810.

[65] Thomsen T, Schlosser A, Holmskov U, Sorensen GL. Ficolins and FIBCD1: soluble and membrane bound pattern recognition molecules with acetyl group selectivity. Mol Immunol 2011;48(4):369–81. https://doi.org/10.1016/j.molimm.2010.09.019. PubMed PMID: 21071088.

[66] Tsuji S, Uehori J, Matsumoto M, Suzuki Y, Matsuhisa A, Toyoshima K, et al. Human intelectin is a novel soluble lectin that recognizes galactofuranose in carbohydrate chains of bacterial cell wall. J Biol Chem 2001;276(26):23456–63. https://doi.org/10.1074/jbc. M103162200. PubMed PMID: 11313366; PubMed Central PMCID: PMCPMC2699772.

[67] Voehringer D, Stanley SA, Cox JS, Completo GC, Lowary TL, Locksley RM. Nippostrongylus brasiliensis: identification of intelectin-1 and -2 as Stat6-dependent genes expressed in lung and intestine during infection. Exp Parasitol 2007;116(4):458–66. https://doi.org/10.1016/j.exppara.2007.02.015. PubMed PMID: 17420014.

[68] O'Neil DA, Porter EM, Elewaut D, Anderson GM, Eckmann L, Ganz T, et al. Expression and regulation of the human beta-defensins hBD-1 and hBD-2 in intestinal epithelium. J Immunol 1999;163(12):6718–24. PubMed PMID: 10586069.

[69] Fahlgren A, Hammarstrom S, Danielsson A, Hammarstrom ML. Increased expression of antimicrobial peptides and lysozyme in colonic epithelial cells of patients with ulcerative colitis. Clinical and experimental immunology. 131(1):90–101. PubMed PMID: 12519391; PubMed Central PMCID: PMCPMC1808590.

[70] Wehkamp J, Schwind B, Herrlinger KR, Baxmann S, Schmidt K, Duchrow M, et al. Innate immunity and colonic inflammation: enhanced expression of epithelial alpha-defensins. Dig Dis Sci 2002;47(6):1349–55. PubMed PMID: 12064812.

[71] Scheetz T, Bartlett JA, Walters JD, Schutte BC, Casavant TL, McCray Jr PB. Genomics-based approaches to gene discovery in innate immunity. Immunol Rev 2002;190:137–45. PubMed PMID: 12493011.

[72] Zanetti M. Cathelicidins, multifunctional peptides of the innate immunity. J Leukoc Biol 2004;75(1):39–48. https://doi.org/10.1189/jlb.0403147. PubMed PMID: 12960280.

[73] Malm J, Sorensen O, Persson T, Frohm-Nilsson M, Johansson B, Bjartell A, et al. The human cationic antimicrobial protein (hCAP-18) is expressed in the epithelium of human epididymis, is present in seminal plasma at high concentrations, and is attached to spermatozoa.

Infection and immunity. 68(7):4297–302. PubMed PMID: 10858248; PubMed Central PMCID: PMCPMC101750.

[74] Mendez-Samperio P. The human cathelicidin hCAP18/LL-37: a multifunctional peptide involved in mycobacterial infections. Peptides 2010;31(9):1791–8. https://doi.org/10.1016/j.peptides.2010.06.016. PubMed PMID: 20600427.

[75] Durr UH, Sudheendra US, Ramamoorthy A. LL-37, the only human member of the cathelicidin family of antimicrobial peptides. Biochim Biophys Acta 2006;1758(9):1408–25. https://doi.org/10.1016/j.bbamem.2006.03.030. PubMed PMID: 16716248.

[76] Eckmann L. Defence molecules in intestinal innate immunity against bacterial infections. Curr Opin Gastroenterol 2005;21(2):147–51. PubMed PMID: 15711205.

[77] Ganz T. Iron in innate immunity: starve the invaders. Curr Opin Immunol 2009;21(1):63–7. https://doi.org/10.1016/j.coi.2009.01.011. PubMed PMID: 19231148; PubMed Central PMCID: PMCPMC2668730.

[78] Raffatellu M, George MD, Akiyama Y, Hornsby MJ, Nuccio SP, Paixao TA, et al. Lipocalin-2 resistance confers an advantage to Salmonella enterica serotype Typhimurium for growth and survival in the inflamed intestine. Cell Host Microbe 2009;5(5):476–86. https://doi.org/10.1016/j.chom.2009.03.011. PubMed PMID: 19454351; PubMed Central PMCID: PMCPMC2768556.

[79] Flo TH, Smith KD, Sato S, Rodriguez DJ, Holmes MA, Strong RK, et al. Lipocalin 2 mediates an innate immune response to bacterial infection by sequestrating iron. Nature 2004;432 (7019):917–21. https://doi.org/10.1038/nature03104. PubMed PMID: 15531878.

[80] Sunil VR, Patel KJ, Nilsen-Hamilton M, Heck DE, Laskin JD, Laskin DL. Acute endotoxemia is associated with upregulation of lipocalin 24p3/Lcn2 in lung and liver. Exp Mol Pathol 2007;83(2):177–87. https://doi.org/10.1016/j.yexmp.2007.03.004. PubMed PMID: 17490638; PubMed Central PMCID: PMCPMC3954125.

[81] Hill CP, Yee J, Selsted ME, Eisenberg D. Crystal structure of defensin HNP-3, an amphiphilic dimer: mechanisms of membrane permeabilization. Science (New York, NY) 1991;251 (5000):1481–5. PubMed PMID: 2006422.

[82] Zhang Y, Lu W, Hong M. The membrane-bound structure and topology of a human alphadefensin indicate a dimer pore mechanism for membrane disruption. Biochemistry 2010; 49(45):9770–82. https://doi.org/10.1021/bi101512j. PubMed PMID: 20961099; PubMed Central PMCID: PMCPMC2992833.

[83] Nash S, Stafford J, Madara JL. Effects of polymorphonuclear leukocyte transmigration on the barrier function of cultured intestinal epithelial monolayers. J Clin Invest 1987;80 (4):1104–13. https://doi.org/10.1172/JCI113167. PubMed PMID: 3116044; PubMed Central PMCID: PMCPMC442353.

[84] Nusrat A, Parkos CA, Liang TW, Carnes DK, Madara JL. Neutrophil migration across model intestinal epithelia: monolayer disruption and subsequent events in epithelial repair. Gastroenterology 1997;113(5):1489–500. PubMed PMID: 9352851.

[85] Kucharzik T, Walsh SV, Chen J, Parkos CA, Nusrat A. Neutrophil transmigration in inflammatory bowel disease is associated with differential expression of epithelial intercellular junction proteins. Am J Pathol 2001;159(6):2001–9. https://doi.org/10.1016/S0002-9440 (10)63051-9. PubMed PMID: 11733350; PubMed Central PMCID: PMCPMC1850599.

[86] Fournier BM, Parkos CA. The role of neutrophils during intestinal inflammation. Mucosal Immunol 2012;5(4):354–66. https://doi.org/10.1038/mi.2012.24. PubMed PMID: 22491176.

[87] O'Shea EF, Cotter PD, Stanton C, Ross RP, Hill C. Production of bioactive substances by intestinal bacteria as a basis for explaining probiotic mechanisms: bacteriocins and conjugated linoleic acid. Int J Food Microbiol 2012;152(3):189–205. https://doi.org/10.1016/j.ijfoodmicro.2011.05.025. PubMed PMID: 21742394.

[88] Nishie M, Nagao J, Sonomoto K. Antibacterial peptides "bacteriocins": an overview of their diverse characteristics and applications. Biocontrol Sci 2012;17(1):1–16. PubMed PMID: 22451427.

[89] Duquesne S, Destoumieux-Garzon D, Peduzzi J, Rebuffat S. Microcins, gene-encoded antibacterial peptides from enterobacteria. Nat Prod Rep 2007;24(4):708–34. https://doi.org/10.1039/b516237h. PubMed PMID: 17653356.

[90] Klaenhammer TR. Genetics of bacteriocins produced by lactic acid bacteria. FEMS Microbiol Rev 1993;12(1–3):39–85. PubMed PMID: 8398217.

[91] Reid G, Burton J. Use of Lactobacillus to prevent infection by pathogenic bacteria. Microbes Infect 2002;4(3):319–24. PubMed PMID: 11909742.

[92] Fuller R. Probiotics in human medicine. Gut 1991;32(4):439–42. PubMed PMID: 1902810; PubMed Central PMCID: PMCPMC1379087.

[93] Cotter PD, Hill C, Ross RP. Bacteriocins: developing innate immunity for food. Nat Rev 2005;3(10):777–88. https://doi.org/10.1038/nrmicro1273. PubMed PMID: 16205711.

[94] Rausch JM, Marks JR, Rathinakumar R, Wimley WC. Beta-sheet pore-forming peptides selected from a rational combinatorial library: mechanism of pore formation in lipid vesicles and activity in biological membranes. Biochemistry 2007;46(43):12124–39. https://doi.org/10.1021/bi700978h. PubMed PMID: 17918962; PubMed Central PMCID: PMCPMC2583027.

[95] Hilpert K, Volkmer-Engert R, Walter T, Hancock RE. High-throughput generation of small antibacterial peptides with improved activity. Nat Biotechnol 2005;23(8):1008–12. https://doi.org/10.1038/nbt1113. PubMed PMID: 16041366.

[96] Wimley WC. Describing the mechanism of antimicrobial peptide action with the interfacial activity model. ACS Chem Biol 2010;5(10):905–17. https://doi.org/10.1021/cb1001558. PubMed PMID: 20698568; PubMed Central PMCID: PMCPMC2955829.

[97] Lichtenstein AK, Ganz T, Nguyen TM, Selsted ME, Lehrer RI. Mechanism of target cytolysis by peptide defensins. Target cell metabolic activities, possibly involving endocytosis, are crucial for expression of cytotoxicity. J Immunol 1988;140(8):2686–94. PubMed PMID: 3162745.

[98] Yeaman MR, Yount NY. Mechanisms of antimicrobial peptide action and resistance. Pharmacol Rev 2003;55(1):27–55. https://doi.org/10.1124/pr.55.1.2. PubMed PMID: 12615953.

[99] Vega LA, Caparon MG. Cationic antimicrobial peptides disrupt the Streptococcus pyogenes ExPortal. Mol Microbiol 2012;85(6):1119–32. https://doi.org/10.1111/j.1365-2958.2012.08163.x. PubMed PMID: 22780862; PubMed Central PMCID: PMCPMC3646575.

[100] Matsuzaki K, Sugishita K, Harada M, Fujii N, Miyajima K. Interactions of an antimicrobial peptide, magainin 2, with outer and inner membranes of Gram-negative bacteria. Biochim Biophys Acta 1997;1327(1):119–30. PubMed PMID: 9247173.

[101] Rathinakumar R, Walkenhorst WF, Wimley WC. Broad-spectrum antimicrobial peptides by rational combinatorial design and high-throughput screening: the importance of interfacial activity. J Am Chem Soc 2009;131(22):7609–17. https://doi.org/10.1021/ja8093247. PubMed PMID: 19445503; PubMed Central PMCID: PMCPMC2935846.

[102] Huang HW. Action of antimicrobial peptides: two-state model. Biochemistry 2000;39(29):8347–52. PubMed PMID: 10913240.

[103] Chen FY, Lee MT, Huang HW. Evidence for membrane thinning effect as the mechanism for peptide-induced pore formation. Biophys J 2003;84(6):3751–8. https://doi.org/10.1016/S0006-3495(03)75103-0. PubMed PMID: 12770881; Central PMCID: PMCPM.

[104] Friedrich CL, Moyles D, Beveridge TJ, Hancock RE. Antibacterial action of structurally diverse cationic peptides on gram-positive bacteria. Antimicrob Agents Chemother 2000;44(8):2086–92. Epub 2000/07/18. PubMed PMID: 10898680; PubMed Central PMCID: PMCPMC1302957.

[105] Yang L, Harroun TA, Weiss TM, Ding L, Huang HW. Barrel-stave model or toroidal model? A case study on melittin pores. Biophys J 2001;81(3):1475–85. https://doi.org/10.1016/S0006-3495(01)75802-X. PubMed PMID: 11509361; PubMed Central PMCID: PMCPMC1301626.

[106] Matsuzaki K, Fukui M, Fujii N, Miyajima K. Interactions of an antimicrobial peptide, tachyplesin I, with lipid membranes. Biochim Biophys Acta 1991;1070(1):259–64. PubMed PMID: 1751532.

[107] Mor A, Nicolas P. The NH2-terminal alpha-helical domain 1-18 of dermaseptin is responsible for antimicrobial activity. J Biol Chem 1994;269(3):1934–9. PubMed PMID: 8294443.

[108] Oren Z, Shai Y. Mode of action of linear amphipathic alpha-helical antimicrobial peptides. Biopolymers 1998;47(6):451–63. https://doi.org/10.1002/(SICI)1097-0282(1998)47:6<451::AID-BIP4>3.0.CO;2-F. PubMed PMID: 10333737.

[109] Dathe M, Wieprecht T. Structural features of helical antimicrobial peptides: their potential to modulate activity on model membranes and biological cells. Biochim Biophys Acta 1999;1462(1–2):71–87. PubMed PMID: 10590303.

[110] Matsuzaki K, Sugishita K, Ishibe N, Ueha M, Nakata S, Miyajima K, et al. Relationship of membrane curvature to the formation of pores by magainin 2. Biochemistry 1998;37(34):11856–63. https://doi.org/10.1021/bi980539y. PubMed PMID: 9718308.

[111] Dathe M, Meyer J, Beyermann M, Maul B, Hoischen C, Bienert M. General aspects of peptide selectivity towards lipid bilayers and cell membranes studied by variation of the structural parameters of amphipathic helical model peptides. Biochim Biophys Acta 2002;1558(2):171–86. PubMed PMID: 11779567.

[112] Duclohier H. Anion pores from magainins and related defensive peptides. Toxicology 1994;87(1-3):175–88. PubMed PMID: 7512759.

[113] Ganz T. Defensins: antimicrobial peptides of vertebrates. C R Biol 2004;327(6):539–49. PubMed PMID: 15330253.

[114] Nevalainen TJ, Graham GG, Scott KF. Antibacterial actions of secreted phospholipases A2. Review. Biochim Biophys Acta 2008;1781(1-2):1–9. https://doi.org/10.1016/j.bbalip.2007.12.001. PubMed PMID: 18177747.

[115] Koprivnjak T, Peschel A. Bacterial resistance mechanisms against host defense peptides. Cell Mol Life Sci 2011;68(13):2243–54. Epub 2011/05/12. https://doi.org/10.1007/s00018-011-0716-4. PubMed PMID: 21560069.

[116] Harwig SS, Tan L, Qu XD, Cho Y, Eisenhauer PB, Lehrer RI. Bactericidal properties of murine intestinal phospholipase A2. J Clin Invest 1995;95(2):603–10. https://doi.org/10.1172/JCI117704. PubMed PMID: 7860744; PubMed Central PMCID: PMCPMC295524.

[117] Lehotzky RE, Partch CL, Mukherjee S, Cash HL, Goldman WE, Gardner KH, et al. Molecular basis for peptidoglycan recognition by a bactericidal lectin. Proc Natl Acad Sci U S A 2010;107(17):7722–7. https://doi.org/10.1073/pnas.0909449107. PubMed PMID: 20382864; PubMed Central PMCID: PMCPMC2867859.

[118] Loonen LM, Stolte EH, Jaklofsky MT, Meijerink M, Dekker J, van Baarlen P, et al. REG3gamma-deficient mice have altered mucus distribution and increased mucosal inflammatory responses to the microbiota and enteric pathogens in the ileum. Mucosal Immunol 2014;7(4):939–47. https://doi.org/10.1038/mi.2013.109. PubMed PMID: 24345802.

[119] Hong RW, Shchepetov M, Weiser JN, Axelsen PH. Transcriptional profile of the Escherichia coli response to the antimicrobial insect peptide cecropin A. Antimicrob Agents Chemother 2003;47(1):1–6. PubMed PMID: 12499161; PubMed Central PMCID: PMCPMC149021.

[120] Papo N, Shai Y. Can we predict biological activity of antimicrobial peptides from their interactions with model phospholipid membranes?. Peptides 2003;24(11):1693–703. https://doi.org/10.1016/j.peptides.2003.09.013. PubMed PMID: 15019200.

[121] Matsuzaki K, Sugishita K, Miyajima K. Interactions of an antimicrobial peptide, magainin 2, with lipopolysaccharide-containing liposomes as a model for outer membranes of gram-negative bacteria. FEBS Lett 1999;449(2–3):221–4. PubMed PMID: 10338136.

[122] Kandaswamy K, Liew TH, Wang CY, Huston-Warren E, Meyer-Hoffert U, Hultenby K, et al. Focal targeting by human beta-defensin 2 disrupts localized virulence factor assembly sites in Enterococcus faecalis. Proc Natl Acad Sci U S A 2013;110(50):20230–5. https://doi.org/10.1073/pnas.1319066110. PubMed PMID: 24191013; PubMed Central PMCID: PMCPMC3864318.

[123] Shi J, Ross CR, Chengappa MM, Sylte MJ, McVey DS, Blecha F. Antibacterial activity of a synthetic peptide (PR-26) derived from PR-39, a proline-arginine-rich neutrophil antimicrobial peptide. Antimicrob Agents Chemother 1996;40(1):115–21. PubMed PMID: 8787891; PubMed Central PMCID: PMCPMC163068.

[124] Chileveru HR, Lim SA, Chairatana P, Wommack AJ, Chiang IL, Nolan EM. Visualizing attack of Escherichia coli by the antimicrobial peptide human defensin 5. Biochemistry 2015;54(9):1767–77. https://doi.org/10.1021/bi501483q. PubMed PMID: 25664683; PubMed Central PMCID: PMCPMC5270551.

[125] Lehrer RI, Barton A, Daher KA, Harwig SS, Ganz T, Selsted ME. Interaction of human defensins with Escherichia coli. Mechanism of bactericidal activity. J Clin Invest 1989;84 (2):553–61. Epub 1989/08/01. https://doi.org/10.1172/JCI114198. PubMed PMID: 2668334; PubMed Central PMCID: PMC548915.

[126] Boman HG, Agerberth B, Boman A. Mechanisms of action on Escherichia coli of cecropin P1 and PR-39, two antibacterial peptides from pig intestine. Infect Immun 1993;61(7):2978–84. PubMed PMID: 8514403; PubMed Central PMCID: PMCPMC280948.

[127] Subbalakshmi C, Sitaram N. Mechanism of antimicrobial action of indolicidin. FEMS Microbiol Lett 1998;160(1):91–6. PubMed PMID: 9495018.

[128] de Leeuw E, Li C, Zeng P, Diepeveen-de Buin M, Lu WY, Breukink E, et al. Functional interaction of human neutrophil peptide-1 with the cell wall precursor lipid II. FEBS Lett 2010;584(8):1543–8. Epub 2010/03/11. https://doi.org/10.1016/j.febslet.2010.03.004. PubMed PMID: 20214904; PubMed Central PMCID: PMCPMC280948.

[129] Sass V, Schneider T, Wilmes M, Korner C, Tossi A, Novikova N, et al. Human beta-defensin 3 inhibits cell wall biosynthesis in Staphylococci. Infect Immun 2010;78(6):2793–800. https://doi.org/10.1128/IAI.00688-09. PubMed PMID: 20385753; PubMed Central PMCID: PMCPMC2876548.

[130] Wilmes M, Sahl HG. Defensin-based anti-infective strategies. Int J Med Microbiol 2014; 304(1):93–9. https://doi.org/10.1016/j.ijmm.2013.08.007. PubMed PMID: 24119539.

[131] Mathew B, Nagaraj R. Variations in the interaction of human defensins with Escherichia coli: possible implications in bacterial killing. PLoS One 2017;12(4). https://doi.org/10.1371/journal.pone.0175858 PubMed PMID: 28423004; PubMed Central PMCID: PMCPMC5397029.

[132] Chairatana P, Nolan EM. Molecular basis for self-assembly of a human host-defense peptide that entraps bacterial pathogens. J Am Chem Soc 2014;136(38):13267–76. https://doi.org/10.1021/ja5057906. PubMed PMID: 25158166; PubMed Central PMCID: PMCPMC4183631.

[133] Chairatana P, Chiang IL, Nolan EM. Human alpha-defensin 6 self-assembly prevents adhesion and suppresses virulence traits of Candida albicans. Biochemistry 2017;56(8):1033–41. https://doi.org/10.1021/acs.biochem.6b01111. PubMed PMID: 28026958.

[134] Zhao A, Lu W, de Leeuw E. Functional synergism of human defensin 5 and human defensin 6. Biochem Biophys Res Commun 2015;467(4):967–72. https://doi.org/10.1016/j.bbrc.2015.10.035. PubMed PMID: 26474700.

[135] Chairatana P, Nolan EM. Defensins, lectins, mucins, and secretory immunoglobulin A: microbe-binding biomolecules that contribute to mucosal immunity in the human gut. Crit Rev Biochem Mol Biol 2017;52(1):45–56. https://doi.org/10.1080/10409238.2016.1243654. PubMed PMID: 27841019; PubMed Central PMCID: PMCPMC5233583.

[136] Mukherjee S, Hooper LV. Antimicrobial defense of the intestine. Immunity 2015;42(1): 28–39. https://doi.org/10.1016/j.immuni.2014.12α.028. PubMed PMID: 25607457.

[137] Lehrer RI, Jung G, Ruchala P, Andre S, Gabius HJ, Lu W. Multivalent binding of carbohydrates by the human alpha-defensin, HD5. J Immunol 2009;183(1):480–90. Epub 2009/06/23. https://doi.org/183/1/480. https://doi.org/[pii]10.4049/jimmunol.0900244. PubMed PMID: 19542459.

[138] Kim C, Gajendran N, Mittrucker HW, Weiwad M, Song YH, Hurwitz R, et al. Human alpha-defensins neutralize anthrax lethal toxin and protect against its fatal consequences. Proc Natl Acad Sci U S A 2005;102(13):4830–5. Epub 2005/03/18. https://doi.org/10.1073/pnas.0500508102. PubMed PMID: 15772169; PubMed Central PMCID: PMC555714.

[139] Giesemann T, Guttenberg G, Aktories K. Human alpha-defensins inhibit Clostridium difficile toxin B. Gastroenterology 2008;134(7):2049–58. PubMed PMID: 18435932.

[140] Wang C, Shen M, Zhang N, Wang S, Xu Y, Chen S, et al. Reduction impairs the antibacterial activity but benefits the LPS neutralization ability of human enteric defensin 5. Sci Rep 2016;6. https://doi.org/10.1038/srep22875 PubMed PMID: 26960718; PubMed Central PMCID: PMCPMC4785407.

[141] Kudryashova E, Quintyn R, Seveau S, Lu W, Wysocki VH, Kudryashov DS. Human defensins facilitate local unfolding of thermodynamically unstable regions of bacterial protein toxins. Immunity 2014;41(5):709–21. https://doi.org/10.1016/j.immuni.2014.10.018. PubMed PMID: 25517613; PubMed Central PMCID: PMCPMC4269836.

[142] Henzler Wildman KA, Lee DK, Ramamoorthy A. Mechanism of lipid bilayer disruption by the human antimicrobial peptide, LL-37. Biochemistry 2003;42(21):6545–58. https://doi.org/10.1021/bi0273563. PubMed PMID: 12767238.

[143] Singh D, Vaughan R, Kao CC. LL-37 peptide enhancement of signal transduction by Toll-like receptor 3 is regulated by pH: identification of a peptide antagonist of LL-37. J Biol Chem 2014;289(40):27614–24. https://doi.org/10.1074/jbc.M114.582973. PubMed PMID: 25092290; PubMed Central PMCID: PMCPMC4183800.

[144] Hassan M, Kjos M, Nes IF, Diep DB, Lotfipour F. Natural antimicrobial peptides from bacteria: characteristics and potential applications to fight against antibiotic resistance. J Appl Microbiol 2012;113(4):723–36. https://doi.org/10.1111/j.1365-2672.2012.05338.x. PubMed PMID: 22583565.

[145] Dobson A, Cotter PD, Ross RP, Hill C. Bacteriocin production: a probiotic trait? Appl Environ Microbiol 2012;78(1):1–6. https://doi.org/10.1128/AEM.05576-11. PubMed PMID: 22038602; PubMed Central PMCID: PMCPMC3255625.

[146] Chikindas ML, Weeks R, Drider D, Chistyakov VA, Dicks LM. Functions and emerging applications of bacteriocins. Curr Opin Biotechnol 2017;49:23–8. 10.1016/j.copbio.2017.07.011. PubMed PMID: 28787641.

[147] Brogden KA, Ackermann M, McCray Jr PB, Tack BF. Antimicrobial peptides in animals and their role in host defences. Int J Antimicrob Agents 2003;22(5):465–78. PubMed PMID: 14602364.

[148] Goldman MJ, Anderson GM, Stolzenberg ED, Kari UP, Zasloff M, Wilson JM. Human beta-defensin-1 is a salt-sensitive antibiotic in lung that is inactivated in cystic fibrosis. Cell 1997;88(4):553–60. Epub 1997/02/21. PubMed PMID: 9038346.

[149] Bastos P, Trindade F, da Costa J, Ferreira R, Vitorino R. Human antimicrobial peptides in bodily fluids: current knowledge and therapeutic perspectives in the postantibiotic era. Med Res Rev 2017;38(1):101–46. https://doi.org/10.1002/med.21435. PubMed PMID: 28094448.

[150] Harder J, Bartels J, Christophers E, Schroder JM. Isolation and characterization of human beta -defensin-3, a novel human inducible peptide antibiotic. J Biol Chem 2001;276(8): 5707–13. Epub 2000/11/22. https://doi.org/10.1074/jbc.M008557200. PubMed PMID: 11085990.

[151] Wilson CL, Ouellette AJ, Satchell DP, Ayabe T, Lopez-Boado YS, Stratman JL, et al. Regulation of intestinal alpha-defensin activation by the metalloproteinase matrilysin in innate host defense. Science (New York, NY) 1999;286(5437):113–7. PubMed PMID: 10506557.

[152] Dubberke ER, Olsen MA. Burden of Clostridium difficile on the healthcare system. Clin Infect Dis 2012;55(Suppl 2):S88–92. Epub 2012/07/07. https://doi.org/10.1093/cid/cis335 PubMed PMID: 22752870; PubMed Central PMCID: PMC3388018..

[153] Ananthakrishnan AN. Clostridium difficile infection: epidemiology, risk factors and management. Nat Rev Gastroenterol Hepatol 2011;8(1):17–26. Epub 2010/12/02. https://doi.org/10.1038/nrgastro.2010.190. PubMed PMID: 21119612.

[154] Kachrimanidou M, Malisiovas N. Clostridium difficile infection: a comprehensive review. Crit Rev Microbiol 2011;37(3):178–87. https://doi.org/10.3109/1040841X.2011.556598. PubMed PMID: 21609252.

[155] Chang JY, Antonopoulos DA, Kalra A, Tonelli A, Khalife WT, Schmidt TM, et al. Decreased diversity of the fecal Microbiome in recurrent Clostridium difficile-associated diarrhea. J Infect Dis 2008;197(3):435–8. Epub 2008/01/18. https://doi.org/10.1086/525047 PubMed PMID: 18199029.

[156] Madan R, Jr WA. Immune responses to Clostridium difficile infection. Trends Mol Med 2012;18(11):658–66. Epub 2012/10/23. https://doi.org/10.1016/j.molmed.2012.09.005 PubMed PMID: 23084763; PubMed Central PMCID: PMC3500589.

[157] Spigaglia P, Barbanti F, Dionisi AM, Mastrantonio P. Clostridium difficile isolates resistant to fluoroquinolones in Italy: emergence of PCR ribotype 018. J Clin Microbiol 2010;48 (8):2892–6. Epub 2010/06/18. https://doi.org/10.1128/JCM.02482-09 PubMed PMID: 20554809; PubMed Central PMCID: PMC2916588.

[158] Walker AS, Eyre DW, Wyllie DH, Dingle KE, Griffiths D, Shine B, et al. Relationship between bacterial strain type, host biomarkers, and mortality in Clostridium difficile infection. Clin Infect Dis 2013;56(11):1589–600. Epub 2013/03/07. https://doi.org/10.1093/cid/cit127 PubMed PMID: 23463640; PubMed Central PMCID: PMC3641870.

[159] Bauer MP, Notermans DW, van Benthem BH, Brazier JS, Wilcox MH, Rupnik M, et al. Clostridium difficile infection in Europe: a hospital-based survey. Lancet 2011;377 (9759):63–73. Epub 2010/11/19. https://doi.org/10.1016/S0140-6736(10)61266-4 PubMed PMID: 21084111.

[160] Tenover FC, Akerlund T, Gerding DN, Goering RV, Bostrom T, Jonsson AM, et al. Comparison of strain typing results for Clostridium difficile isolates from North America. J Clin Microbiol 2011;49(5):1831–7. Epub 2011/03/11. https://doi.org/10.1128/JCM.02446-10. PubMed PMID: 21389155; PubMed Central PMCID: PMC3122689.

[161] Vedantam G, Clark A, Chu M, McQuade R, Mallozzi M, Viswanathan VK. Clostridium difficile infection: toxins and non-toxin virulence factors, and their contributions to disease establishment and host response. Gut Microbes 2012;3(2):121–34. Epub 2012/05/05. https://doi.org/10.4161/gmic.19399. PubMed PMID: 22555464; PubMed Central PMCID: PMC3370945.

[162] McBride SM, Sonenshein AL. The dlt operon confers resistance to cationic antimicrobial peptides in Clostridium difficile. Microbiology 2011;157(Pt 5):1457–65. Epub 2011/02/ 19. https://doi.org/10.1099/mic.0.045997-0. PubMed PMID: 21330441; PubMed Central PMCID: PMC3140582.

[163] McBride SM, Sonenshein AL. Identification of a genetic locus responsible for antimicrobial peptide resistance in Clostridium difficile. Infect Immun 2011;79(1):167–76. Epub 2010/10/ 27. https://doi.org/10.1128/IAI.00731-10 PubMed PMID: 20974818.

[164] Ho TD, Ellermeier CD. PrsW is required for colonization, resistance to antimicrobial peptides, and expression of extracytoplasmic function sigma factors in Clostridium difficile. Infect Immun 2011;79(8):3229–38. Epub 2011/06/02. https://doi.org/10.1128/IAI.00019-11 PubMed PMID: 21628514.

[165] McQuade R, Roxas B, Viswanathan VK, Vedantam G. Clostridium difficile clinical isolates exhibit variable susceptibility and proteome alterations upon exposure to mammalian cationic antimicrobial peptides. Anaerobe 2012;18(6):614–20. Epub 2012/09/29. https:// doi.org/10.1016/j.anaerobe.2012.09.004. PubMed PMID: 23017940.

[166] Hasegawa M, Yamazaki T, Kamada N, Tawaratsumida K, Kim YG, Nunez G, et al. Nucleotide-binding oligomerization domain 1 mediates recognition of Clostridium difficile and induces neutrophil recruitment and protection against the pathogen. J Immunol 2011;186 (8):4872–80. Epub 2011/03/18. https://doi.org/10.4049/jimmunol.1003761. PubMed PMID: 21411735.

[167] Jarchum I, Liu M, Shi C, Equinda M, Pamer EG. Critical role for MyD88-mediated neutrophil recruitment during Clostridium difficile colitis. Infect Immun 2012;80(9):2989–96. Epub 2012/06/13. https://doi.org/10.1128/IAI.00448-12. PubMed PMID: 22689818.

[168] Peschel A, Sahl HG. The co-evolution of host cationic antimicrobial peptides and microbial resistance. Nat Rev Microbiol 2006;4(7):529–36. Epub 2006/06/17. https://doi.org/10.1038/ nrmicro1441. PubMed PMID: 16778838.

[169] Huang HW. Free energies of molecular bound states in lipid bilayers: lethal concentrations of antimicrobial peptides. Biophys J 2009;96(8):3263–72. Epub 2009/04/23. https://doi.org/ 10.1016/j.bpj.2009.01.030. PubMed PMID: 19383470; PubMed Central PMCID: PMC2718316.

[170] Shimoda M, Ohki K, Shimamoto Y, Kohashi O. Morphology of defensin-treated Staphylococcus aureus. Infect Immun 1995;63(8):2886–91. Epub 1995/08/01. PubMed PMID: 7622209; PubMed Central PMCID: PMC173392.

[171] Hartmann M, Berditsch M, Hawecker J, Ardakani MF, Gerthsen D, Ulrich AS. Damage of the bacterial cell envelope by antimicrobial peptides gramicidin S and PGLa as revealed by transmission and scanning electron microscopy. Antimicrob Agents Chemother 2010; 54(8):3132–42. Epub 2010/06/10. https://doi.org/10.1128/AAC.00124-10. PubMed PMID: 20530225; PubMed Central PMCID: PMC2916356.

[172] Rajabi M, Ericksen B, Wu X, de Leeuw E, Zhao L, Pazgier M, et al. Functional determinants of human enteric alpha-defensin HD5: crucial role for hydrophobicity at dimer interface. J Biol Chem 2012;287(26):21615–27. Epub 2012/05/11. https://doi.org/10.1074/jbc.M112. 367995 PubMed PMID: 22573326; PubMed Central PMCID: PMC3381126.

[173] Hiemstra PS, Zaat SA. Antimicrobial peptides and innate immunity. Springer; 2013.

[174] Menendez A, Willing BP, Montero M, Wlodarska M, So CC, Bhinder G, et al. Bacterial stimulation of the TLR-MyD88 pathway modulates the homeostatic expression of ileal Paneth cell alpha-defensins. J Innate Immun 2012;5(1):39–49. Epub 2012/09/19. https://doi.org/ 10.1159/000341630. PubMed PMID: 22986642.

[175] van Nood E, Vrieze A, Nieuwdorp M, Fuentes S, Zoetendal EG, de Vos WM, et al. Duodenal infusion of donor feces for recurrent Clostridium difficile. N Engl J Med 2013;368(5): 407–15. Epub 2013/01/18. https://doi.org/10.1056/NEJMoa1205037. PubMed PMID: 23323867.

[176] Yeung AT, Gellatly SL, Hancock RE. Multifunctional cationic host defence peptides and their clinical applications. Cell Mol Life Sci 2011;68(13):2161–76. Epub 2011/05/17. https://doi.org/10.1007/s00018-011-0710-x. PubMed PMID: 21573784.

[177] Brandl K, Plitas G, Schnabl B, DeMatteo RP, Pamer EG. MyD88-mediated signals induce the bactericidal lectin RegIII gamma and protect mice against intestinal Listeria monocytogenes infection. J Exp Med 2007;204(8):1891–900. https://doi.org/10.1084/jem.20070563. PubMed PMID: 17635956; PubMed Central PMCID: PMCPMC2118673.

[178] Brandl K, Plitas G, Mihu CN, Ubeda C, Jia T, Fleisher M, et al. Vancomycin-resistant enterococci exploit antibiotic-induced innate immune deficits. Nature 2008;455(7214):804–7. https://doi.org/10.1038/nature07250. PubMed PMID: 18724361; PubMed Central PMCID: PMCPMC2663337.

[179] Adolph TE, Tomczak MF, Niederreiter L, Ko HJ, Bock J, Martinez-Naves E, et al. Paneth cells as a site of origin for intestinal inflammation. Nature 2013;503(7475):272–6. https://doi.org/10.1038/nature12599. PubMed PMID: 24089213; PubMed Central PMCID: PMCPMC3862182.

[180] Andersson ML, Karlsson-Sjoberg JM, Putsep KL. CRS-peptides: unique defense peptides of mouse Paneth cells. Mucosal Immunol 2012;5(4):367–76. https://doi.org/10.1038/mi.2012.22. PubMed PMID: 22535181.

[181] Virgin HW. The virome in mammalian physiology and disease. Cell 2014;157(1):142–50. https://doi.org/10.1016/j.cell.2014.02.032. PubMed PMID: 24679532; PubMed Central PMCID: PMCPMC3977141.

[182] Reyes A, Haynes M, Hanson N, Angly FE, Heath AC, Rohwer F, et al. Viruses in the faecal microbiota of monozygotic twins and their mothers. Nature 2010;466(7304):334–8. https://doi.org/10.1038/nature09199. PubMed PMID: 20631792; PubMed Central PMCID: PMCPMC2919852.

[183] Iliev ID, Funari VA, Taylor KD, Nguyen Q, Reyes CN, Strom SP, et al. Interactions between commensal fungi and the C-type lectin receptor Dectin-1 influence colitis. Science (New York, NY) 2012;336(6086):1314–7. https://doi.org/10.1126/science.1221789. PubMed PMID: 22674328; PubMed Central PMCID: PMCPMC3432565.

Chapter 7

Host Defense Peptides as Innate Immunomodulators in the Pathogenesis of Colitis

Ravi Holani*, Maia S. Marin†, John P. Kastelic* and Eduardo R. Cobo*

*University of Calgary, Calgary, Canada, †National Scientific and Technical Research Council (CONICET), Buenos Aires, Argentina

Chapter Outline

1 BACKGROUND

Based on the causative agent, colitis (inflammation of the colon) can broadly be divided into two types: infectious colitis and inflammatory bowel diseases (IBDs). Infectious colitis is caused by a diverse group of pathogens, including bacteria, protozoa, and viruses, such as *Clostridium difficile* (*C. difficile*), *Entamoeba histolytica,* and cytomegalovirus, respectively [1,2]. Most of these infectious agents are transmitted from animals to humans via ingestion of feces or contaminated animal tissues, making them a major food-borne risk. Infectious colitis is the leading cause of diarrhea, morbidity and mortality in children <5 years of age, particularly in developing countries [3]. In the acute stage, infectious colitis is often characterized by excessive fluid loss (diarrhea), abdominal pain, tenesmus, and occasionally rectal bleeding [1]. Although usually self-limiting, in elderly or immunocompromised individuals, infectious colitis can lead to septicemia and death [1].

Antimicrobial Peptides in Gastrointestinal Diseases. https://doi.org/10.1016/B978-0-12-814319-3.00007-6
Copyright © 2018 Chi Hin Cho. Published by Elsevier Ltd. All rights reserved.

IBDs are characterized by chronic relapsing intestinal diseases. Currently, ~3 million people in North America and Europe alone are estimated to have IBD [4]. Although the major incitant(s) for IBDs are not established (idiopathic), this syndrome is often regarded as a multifactorial disorder (genetic predisposition, environmental factors, and immune dysregulation). Based on immunopathology, histopathology, and affected anatomical site(s), IBDs are broadly categorized into two types: Crohn's disease (CD) and ulcerative colitis (UC) [5]. CD is characterized by systemic, discontinous (patchy), and often transmural inflammation of the intestine or other parts of the gastroesophagal system, whereas UC is mostly continous inflammation of the mucosa of the lower colon [5].

Irrespective of the type, colitis is characterized by a damaged/leaky epithelial barrier and dysregulated intestinal immune response. Excessive infiltration of immune cells such as T-cells and lymphoid cells, antigen-presenting cells (APCs; dendritic cells and macrophages), and neutrophils are normally present in an inflamed bowel [6]. Leaky epithelial junctions allow pathogens and other materials to infiltrate the lamina propria. Accumulated pathogenic microbes then induce differentiation and activation of immune cells, resulting in exaggerated inflammatory responses [7,8], perhaps through interaction with pattern recognition receptors (PRRs; innate immune receptors for microbial determinants; [9]). The presence of microbes stimulating excessive inflammatory responses seems to be essential in the pathogenesis of colitis. For instance, gut inflammation is reduced in germ-free mice [10], there are T-cell responses against commensals in IBD patients [11], and antibiotics-reduced clinical signs of CD [12].

Expression of specific cytokines (effector proteins, secreted by a variety of hematopoietic and nonhematopoietic cells), either proinflammatory (e.g., TNFα and IL-6) or antiinflammatory (e.g., IL-10 and TGFβ), are critical in the pathogenesis of colitis [7]. Infectious colitis caused by enteropathogenic *Escherichia coli* (EPEC), *Salmonella typhimurium* (*S. typhimurium*), *Shigella* spp., and *C. difficile* involves overexpression of tumor necrosis factor-alpha (TNF-α), interleukin (IL)-17, IL-6, and IL-1β [13]. In chronic IBDs, whereas CD is characterized by exaggerated production of interferon-γ (IFN-γ), interleukin-12 (IL-12), IL-17, and IL-23 cytokine expression [14,15], UC patients have high expression of IL-13 [16] but peculiarly low expression of IL-4 [17]. The relevance of cytokine expression in colitis is further exemplified by the fact that patients predisposed to chronic colitis often have mutations in cytokine-associated loci [7,18].

Due to the clinical importance of cytokines in IBDs, pharmacological modulation of cytokine expression has been a core therapeutic strategy against colitis for more than two decades [7]. The use of anti-TNF-α antibodies is a current standard therapy for IBD. However, given genetic diversity among individuals and complexity of the cytokine signaling network, therapeutic strategies targeting only one cytokine are often ineffective. In clinical studies, anti-TNF-α therapy was not beneficial at the outset in ~50% of patients; furthermore, in those that initially responded, it eventually became ineffective in approximately 50% [19]. Although anti-TNF-α therapy controls inflammation, it does not prevent or reverse intestinal fibrosis [20], a major reason for surgical intervention in most CD patients [21]. Therefore, novel strategies targeting multiple cytokines such as Janus kinase (JAK) inhibitor (tofacitinib) are appealing and are undergoing clinical trials. Other models of experimental colitis, such as T-cell transfer (chronic) or protein haptenization-based oxa-zolone (acute) and TNBS (chronic), have demonstrated that both immune-suppressive (recombinant IFNβ or anti-TNFα drugs, e.g., infliximab, adalimumab, certolizumab, and golimumab) and immune-stimulatory drugs (e.g., recombinant IL-11) may be necessary to ameliorate exaggerated inflammatory responses in IBDs [7].

2   HOST DEFENSE PEPTIDES

In a search for novel therapeutics for controlling microbial infection and damaging inflammation in the gut, host defense peptides (HDPs), also called antimicrobial peptides (AMPs), have been targeted. HDPs are a family of evolutionarily conserved small cationic amphiphilic peptides with immunomodulatory properties. They are widespread across complex animals, as well as plants and insects. In humans, there are two major classes of HDPs: cathelicidins and defensins, in addition to other less studied AMPs.

LL-37/hCAP-18 is the only cathelicidin expressed in humans, whereas mice also have only one cathelicidin, named cathelicidin-related-antimicrobial-peptide (mCramp) [21a]. In contrast, other animals, such as pigs, express a wide range of cathelicidin peptides [22]. Human cathelicidin has an N-terminal domain containing a signal peptide, a well-conserved central cathelin domain, and a variable C-terminal domain [23]. The 37-amino acid-long C terminus, designated LL-37, represents the antimicrobial-active cationic peptide. These small cationic cathelicidins are abundantly expressed in mammalian cells, mainly neutrophils and epithelial cells [23a]. Thus, cathelicidins are present in a variety of locations, including skin, eyes, mouth, lungs, and intestine [24]. In the gut, epithelial cells and immune cells including neutrophils abundantly express cathelicidins, with constitutive production in intestinal mucosa and induced production in response to epithelial injury [25,26]. Cathelicidins are multifunctional molecules that mediate various host responses, including microbicidal action, chemotaxis, epithelial cell activation, epithelial wound repair, cancer modulation, and activation of chemokine secretion.

Defensins in humans can be classified into two subtypes: α-defensins and β-defensins, based on arrangements of disulfide bonds between six conserved cysteine pairs. There are six α-defensins in humans: neutrophil peptides 1–4 (HNPs), and defensins HD5 and HD6. HNP1–4 defensins are mainly secreted by neutrophils present, for instance, in the lamina propria of the intestine. Patients with either UC or CD have increased secretion of HNP1–3 by their intestinal epithelium in areas of active inflammation [27]. Conversely, HD5 and HD6 are produced as pro-peptides in Paneth cells, specialized intestinal epithelial cells located at the base of crypts in the small intestine. Paneth cells are key effectors of small intestinal antimicrobial defenses and gut homeostasis by regulating composition of the intestinal flora and providing protection against pathogens. Defensin pro-peptides secreted by Paneth cells are cleaved into active forms by endogenous digestive enzyme trypsin or matrix metalloproteinase-7 (MMP-7; mouse homologue) [28]. Paneth cells also contain cytoplasmic secretory granules with other antimicrobial peptides, that is, RegIIIγ [28a,28b]. HD-5 and HD-6 can also be produced in the colon, but are less abundant [28c,29,30]. The other subfamily of defensins is β-defensins, which includes six human β-defensins (hBDs) and five mouse β-defensins (mBDs). In contrast to α-defensins, β-defensins vary in their expression patterns through the gut and are expressed by several classes of intestinal cells, including enterocytes [30a].

2.1 HDPs in Infectious Colitis

Both cathelicidins and defensins are crucial determinants in pathogenesis of infectious colitis; consequently, counter-attack mechanisms are often employed by enteric pathogens to protect themselves against HDPs. For example, a virulent strain of *Shigella flexneri* (*S. flexneri*) reduced expression of LL-37 and hBD-3 mRNA in colonic epithelial (TC7) cells [31]. This evasion mechanism was consistent with studies in human intestinal xenografts in which deep intestinal crypt invasion by *S. flexneri* reduced hBD-1 and -3 protein expression [31]. Likewise, *Cryptosporidium parvum* (*C. parvum*), a protozoan parasite, down-regulates expression of mouse beta-defensin (mBD)-1 and mBD3 peptides in rectal (CMT-93) carcinoma cells, peptides that would otherwise kill *C. parvum* sporozoites [32]. Furthemore, oral challenge of mice with *S. typhimurium* reduced expression of cryptidins (murine homologue of α-defensins) by Paneth cells in a type III secretion system (T3SS)-dependent manner [33]. Importantly, mice deficient in cryptidins had increased susceptibility to *S. typhimurium* infection [28].

Given the importance of HDPs in the innate response to intestinal pathogens, it is not surprising that therapeutic interventions with HDPs have been effective against infectious colitis. For example, intraperitoneal injection of cathelicidin-WA (from *Bungarus fascia*, a venomous snake) mitigated etiologically undefined diarrhea and improved tight junctions and small intestinal epithelial morphology (i.e., increased villus height) in piglets [34]. In addition, lentivirus-mediated overexpression of cathelicidin prevented intestinal fibrosis in C57BL/6J mice orally challenged with *S. typhimurium* [35]. Likewise, intracolonic administration of mCramp prevented inflammation (reduced expression of TNF-α and tissue-specific reductions in neutrophil activity), intestinal epithelial damage, and apopotosis in C57BL/6 mice infected with *C. difficile* [36].

Furthermore, Paneth cell-specific HD-5 overexpression in mice promoted resistance to oral infection with *S. typhimurium* [37]. Interestingly, *Giardia muris* coinfection mitigated weight loss and gross pathology induced by *Citrobacter rodentium* in mice; this protection was correlated with enhanced secretion of antimicrobial proteins, including mBD-3 [38].

2.2   HDPs in IBDs

Recent studies have established an association between HDPs and IBDs. Patients with either UC or CD had increased expression of LL-37 in their serum compared to healthy individuals [39]. At the intestinal level, patients with UC, but not those with CD, had elevated mRNA levels of LL-37 (albeit similar protein expression) in both inflamed and non-inflamed colonic mucosa, compared to noninflamed mucosa of healthy individuals [40]. Similarly, cathelicidin protein expression was not significantly different in colonic biopsies from UC patients compared to healthy individuals [41]. This apparent discrepancy between mRNA and protein levels of cathelicidins in UC patients could be due to either technical limitations to quantify LL-37 protein expression associated with nonquantitative approaches (immunohistochemistry) or innate defects in UC patients (e.g., impaired protein translation, excessive secretion or rapid degradation of the peptide, or limited local cathelicidin synthesis). In this regard, altered distrubution of cathelicidins has been detected in IBDs. Cathelicidins were localized in granules at the base of coloncytes in patients with IBDs, whereas this peptide was diffusely distributed in healthy individuals [40] although the functional relevance is unknown.

Despite the need to better understand the biological role of cathelicidins in patients with IBD, the importance of cathelicidins in gut protection has already been well established in experimental models. In C57BL/6J mice, intracolonic or intravenous administration of lentivirus overexpressing cathelicidins ameliorated gut inflammation and associated fibrosis in response to chemical (TNBS induced colitis) or *S. typhimurium* induced infectious colitis [35]. Likewise, oral treatment in mice with *Lactococcous lactis* expressing mouse cathelicidin-related-antimicrobial-peptide (mCramp) diminished UC-like colitis induced by DSS treatment [42]. Administration of synthetic cathelicidins mCramp into the colon also reduced clinical manifestations of colitis (e.g., body weight and histopathological scores) in response to DSS [43]. Epidemiological approaches in CD patients reinforced therapeutic effects of cathelicidins. Vitamin D treatment (which induces naturally occurring cathelicidins) reduced intestinal permeability and prolonged remission with sustained serum concentrations of 25-hyrdoxycholecalciferol and cathelicidins [44].

Human defensins are also involved in IBDs. Gene expression of HD-5 and HD-6 mRNA was downregulated in ileal CD patients compared to control groups [45] and it was associated with excessive epithelial damage [46]. Furthermore, CD patients have reduced stability of HD-5 peptide, compared to healthy persons [47]. IBDs are also often characterized by mutations in autophagy related 16-like 1 (ATG16L1) and transcription factor-X-box-binding-protein 1 (XBP1) genes, both of which impair Paneth cell function and reduce secretion of α-defensins [48,49]. In agreement, children with CD had reduced expression of a Paneth cell differentiation (Tcf-4) marker in their duodenum and terminal ileum, suggesting deficient Paneth cell development (and α-defensin production) in CD [50].

In the colon, β-defensins are mostly expressed by epithelial cells and plasma cells in the lamina propria. Among the four types of human β-defensins (HBD1–4), HBD1 is constitutively expressed by colonic epithelium, whereas HBD2–4 expression is inducible in response to various inflammatory and bacterial stimuli [51]. Although HBD1 levels remain unaffected in IBD patients, HBD2 mRNA was upregulated in inflamed areas of the intestine in UC and CD patients, compared to control biopsies or biopsies from noninflamed areas of UC and CD patients [52]. Moreover, HBD2 expression in colonic biopsies was significantly higher in UC patients compared to CD patients [52]. Similarly, although HBD3 and HBD4 had minimal expression in healthy controls, colonic biopsies from IBD patients had increased HBD3 expression in villous of CD patients, with both HBD3 and 4 overexpressed in UC patients [53].

Experimental models of colitis in mice have also provided evidence of the protective role of defensins against IBD. Intraperitoneal administration of HD5 ameliorated DSS-induced colitis and enhanced survival compared to untreated mice [54]. A recent patent claims that oral therapy with β-defensins reestablished a normal epithelial barrier in CD patients and thus reduced disease severity [55]. Similarly, HBD1–4 may provide some benefits against IBDs as simultaneous or successive parental and oral administration of β-defensins protected mice against DSS-induced colitis [56]. In agreement, subcutaneous injection of HBD2 (0.1–3 mg/kg) ameliorated colitis induced by DSS, TNBS, or T-cell transfer at similar level compared to conventional therapeutics, for example, anti-TNFα (DSS), prednisone (TNBS), or dexamethasone (transfer) [57]. In summary, despite an incomplete understanding regarding how expression of cathelicidins and defensins are altered in IBD (i.e., cause or consequence), there is consensus regarding a particular dissociation of HDPs in IBDs. The protective nature of HDPs in IBD patients or induced experimental models of colitis in mice highlights the need for further research into functional aspects of HDPs at a molecular level, particularly mechanisms by which HDPs are able to immunomodulate the detrimental gut inflammation that is the hallmark of IBDs.

3 MECHANISMS USED BY HOST DEFENSE PEPTIDES DURING COLITIS

HDPs have important roles in colitis, either incited by biotic or abiotic stressors. In this section, we review how HDPs may promote gut health and resolve colitis, including three main aspects: antibacterial activity (may indirectly resolve inflammation by eliminating the infectious incitant), immunomodulation (via chemotaxis of immune cells and modulation of cytokines), and interactions between HDPs and PRRs (reciprocal regulation in response to microbes/microbial determinants).

3.1   Antibacterial Role of HDPs in the Intestine

IBDs are characterized by the presence of commensal organisms and overgrowth of pathogens, in association with exaggerated gut inflammation. HDPs were initially discovered on the basis of their pronounced antimicrobial action. Although HDP efficacy as direct antimicrobials in the gut remains elusive, in vitro assays have confirmed that these peptides can kill or inhibit microbes by permeabilizing bacterial membranes or inhibiting metabolic pathways [58]. Thus, in vitro, cathelicidins have antimicrobial activity against enteric pathogens such as *E. coli*, *Staphylococcus aureus*, group A *Streptococcus*, and enteroinvasive *E. coli* and *Salmonella* spp. [59,60]. Furthermore, cathelicidins neutralize virulent microbial factors such as LPS and DNA, via direct binding [61,62]. This property of cathelicidins could be pertinent in preventing unwarranted epithelial apoptosis, characteristic of IBDs due to inflammation provoked by microbial components [63]. The role of cathelicidins in ameriolating infection in vivo has also been reported. *Cramp*-deficient mice orally infected with *E. coli* O157:H7 had a four-fold higher microbial burden in their feces than wild type [64], although the role of direct antimicrobial effects was inconclusive.

Regarding defensins, HD5 overexpressing Paneth cells conferred site-specific protective effects in transgenic mice orally challenged by *S. typhimurium* [37]. Likewise, $Mmp-7^{-/-}$ mice (impaired in α-defensin activation) were more susceptible to death by oral inoculation of *S. typhimurium*, with one-third of $Mmp-7^{-/-}$ mice dying within 8 days postinfection, compared to no mortality in wild type [28]. Unlike HD5, inducible HD6 had poor antibacterial activity, attributed to its atypical arrangement of arginine residues [65]. However, under standard intestinal lumen conditions (i.e., pH 7.4 and a reducing environment), HD6 acquired selective antibacterial activity against gut commensals, including *Bifidobacterium breve* and *Streptococcus thermophiles* [66]. The antibacterial activity of constitutively expressed HBD-1 and inducible HBD2–4 has been well documented. Reduced levels of mDefB10 (homologue of human HBD1) in PPAR-γ-deficient mice increased susceptibility to *Candida albicans*, *Bacteroides fragilis*, *Enterococcus faecalis*, and *E. coli* [67]. Defensins also had antimicrobial effects in cultured colonic cells. Gene expression of HBD2 was upregulated in colonic epithelial (HT29) cells in response to *C. parvum* infection; this increase was associated with reduced number, invasion, and infectivity of *C. parvum* sporozoites [32,68].

The antimicrobial activity of HDPs in in vitro systems was dependent on attributes of the killing buffer. Activity of HDPs was reduced at high salt concentrations (100 mM) or in the presence of serum proteins [30, 68a]. For example, killing activity of cathelicidin against group B *Streptococcus* spp. (encapsulated and nonencapsulated) and *Pseudomonas aeruginosa* was reduced ~50% and ~90% in the presence of lung surfactant [69] and artificial tears [70], respectively. Factors in the gut that regulate in vivo killing capacity of these peptides is not well understood. Perhaps specific niches in the gut, for example, bottom of the crypts, can provide optimal conditions for HDP killing activity. Overall, the recognized antibacterial activity of HDPs and their broad bacterial targets make HDPs substantially different from conventional antibiotics, which usually have single mechanisms of action.

3.2 Immunomodulatory Role of HDPs in Colitis

HDPs were initially considered antimicrobial peptides, but it was soon realized that they were more than simply natural antibiotics. Due to their ability to interact with membranes of various origins, HDPs can modulate immune responses through multiple non-antimicrobial mechanisms on host cells. The ability of HDPs to regulate signaling molecules is broad. They can interact with cell surface and intracellular compartments, a plethora of receptors, and a wide variety of immune cells (e.g., macrophages, neutrophils, dendritic cells) and nonimmune cells (e.g., epithelial cells, glial cells, endothelial cells, and fibroblasts; [71]). For example, LL-37 activated epidermal growth factor receptor (EGFR) and p38-mitogen-activated-protein-kinase (MAPK) in colonic epithelial cells (unpublished data from Holani R et al), insulin-like growth factor 1 receptor (IGF1R) and extracellular-signal-regulated-kinases (ERK) in human MCF7 breast cancer cells [72], formyl peptide receptor like 1 (FPRL1/FPR2) in human monocytes [73], interleukin-8 (IL-8) receptor CXCR2 in human neutrophils [74], and an orphan GPCR called mas-related gene ×2 (MrgX2) in human mast cells [75] (Table 1). Similarly, rat cathelicidin-related-antimicrobial-peptide (Cramp) activated a purinergic P2Y11 (GPCR) receptor on glial cells and promoted signaling through ERK1/2 [76] (Table 1 and Fig. 1). Furthermore, hBD3 promoted NFκB signaling in toll-like-receptor 2 (TLR 2) overexpressing kidney epithelial HEK293 cells [77] (Table 1 and Fig. 1). It is noteworthy that all of these receptors and signaling molecules are at the crux of cytokine-mediated immune responses. Whether the cytokine cascade triggered by HDPs ultimately resulted in pro- or antiinflammatory responses depended on cell type and presence of specific inflammatory stimuli, including pathogenic factors and surrounding cytokine milleu. These factors will be discussed in the following sections.

3.2.1 Proinflammatory Chemoattraction and Cell Activation Effects of HDPs

Inflammation (e.g., colitis) involves recruitment of immune cells to the injury site. Both cathelicidins and defensins promote chemotaxis either through direct

TABLE 1 Antimicrobial Peptides and Potential Cellular Receptors

Host Defense Peptide	Involved Cellular Receptor	Full Name Description	Refs.
Cathelicidins	EGFR	Epidermal growth factor receptor	Unpublished data from Holani R and Cobo E
	IGF1R	Insulin-like growth factor receptor 1	[72]
	FPRL1/FPR2	Formyl peptide receptor-like 1/formyl peptide receptor 2	[73]
	CXCR2	Cysteine-X-cysteine receptor 2	[74]
	MrgX2	Mas-related gene × 2	[75]
	P2Y11	Purinoreceptor 11	[76]
	GAPDH	Glyceraldehyde-3-phosphate dehydrogenase	[87]
Defensins	TLR2	Toll-like-receptor 1/2	[77]
	CCR6	Cysteine-cysteine receptor 6	[78]
	CCR2	Cysteine-cysteine receptor 2	[79]
	P2Y6	Purinoreceptor 6	[80]

Although host defense peptides do not possess an specific receptor, several cell receptors have been reported to be activated by either cathelicidins or defenins.

activation of chemotactic receptors or indirectly, via regulation of chemokine expression. Furthermore, human cathelicidin LL-37 promotes in vitro chemotaxis of T-cells, macrophages, neutrophils, eosinophils, and mast cells via interactions with FPRL-1 receptor on their surfaces [73,81] (Table 1 and Fig. 1). In fact, desensitization of FPRL-1 in monocytes using a specific agonist for FPRL-1 receptor (Su peptide, corresponds to amino acids 563–595 of HIV envelope protein gp41) or transfection of HEK293 with FPRL-1 has confirmed this chemotactic role of LL-37 and the importance of FPRL1 [73]. In in vitro studies, murine Cramp attracted monocytes, neutrophils, macrophages, and leukocytes via an FPRL-1 receptor, as demonstrated by competitive inhibition of chemotaxis in the presence of the FPRL-1 agonist MMK-1 [82]. In contrast, LL-37 mediated chemotaxis of mast cells was independent of FPRL-1, but

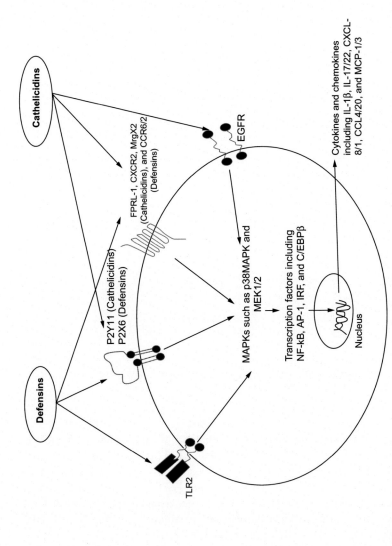

FIG. 1 Schematic representation of the observed intrinsic receptors/signaling mechanisms of action of host defense peptides, cathelicidins, and defensins, in the colonic epithelium.

dependent on Gi-protein-phospholipase C signaling, based on inhibitor assays [83] (Table 1 and Fig. 1).

Other than direct activation of chemokine receptors, LL-37 also induced expression of CXCL8 chemokine (neutrophil chemo attractant) via activation of P2X7 receptor in human gingival fibroblasts [84] and in an EGFR-dependent manner in skin [85] and airway epithelial cells [81]. With regards to leukocytes, cathelicidins induced IL-8 through MEK1/2-p38MAPK activation in human monocytes [86]. Moreover, LL-37 interacted intracellularly with glyceralde-hyde dehydrogenase (GAPDH) in human peripheral blood mononuclear cells (PBMCs) to promote expression of related neutrophil chemoattractants, including CCL-4/MIP-1β, CCL-20/MIP-3α, and CXCL-1/GRO-α [87]. Furthermore, LL-37 stimulated HUVEC endothelial cells, mouse macrophage like RAW264.7, and human peripheral blood cells to produce CCL-2 (also called MCP-1), a chemo attractant for monocytes, T-cells, and dendritic cells [88]. Interestingly, LL-37 synergized with inflammatory stimuli (cytokines or path-ogenic factors) in this enhancement of chemokine expression. Cathelicidin LL-37 enhanced secretion of chemokine monocyte chemo attractant protein (MCP)-1 and MCP-3 in the presence of proinflammatory IL-1β and GM-CSF [89]. Likewise, LL-37 enhanced IL-17 and IL-22 mediated IL-8 response in human keratinocytes [90]. In response to Gram-negative virulent factor LPS (TLR4 agonist), LL-37 stimulated IL-8 secretion in bronchial epi-thelial cells [62]. Moreover, LL-37 in synergy with *S. typhimurium* or LPS induced EGFR dependent IL-8 (and keratinocyte chemo attractant -KC; a murine homologue of IL-8) in intestinal epithelial cells (HT29, T-84 cells), mouse colonoids and in vivo, in *Cramp*$^{-/-}$ mice (unpublished data from Holani R and Cobo E) (Table 1 and Fig. 1).

Defensins regulate chemoattraction of immune cells based on receptor-dependent or -independent (chemokine expression) mechanisms. Unlike cathelicidins, defensins do not promote chemotactic activity via GPCRs like FPRL-1, but through direct interaction with chemokine receptors (Table 1 and Fig. 1). For example, HBDs induced chemotaxis of memory T-cells (CD4+ CD45RO+), CD8+ T cells, and immature dendritic cells (iDCs) through direct interactions with CCR6 chemokine receptors, as confirmed using a natural ligand for CCR6 (LARC/MIP-3α) and anti-CCR6-mediated competitive chemotaxis inhibition assays [78] (Table 1 and Fig. 1). In other studies, HBD1–4 in particular did not attract lymphocytes and dendritic cells via a CCR6 pathway; however, they chemoattracted macrophages and mast cells in a Gi-protein-dependent fash-ion [91], later determined to be CCR2 [79] (Table 1 and Fig. 1). The apparent discrepancy regarding whether HBDs chemoattract lymphocytes and dendritic cells may have been due to a difference in peptide source (one study used HDBs isolated from psoriatic skin that may have included additional factors) [78], whereas the other utilized synthetic peptides only [91]. HNP1–3 has been reported to promote chemotaxis of iDCs, CD8+ T cells and naïve T-cells (CD4+ CD45RA+) but not of memory T-cells [92]. Therefore, unlike HBDs, perhaps HNPs do not signal through the CCR6 receptor. Indeed, it was reported that HNPs signal via a purinergic receptor, named P2Y6 [80] (Table 1 and Fig. 1). Other defensin particularities attracting immune cells have been reported. HNP1–2 but not HNP-3 chemoattracted peripheral blood derived monocytes [93] and apparently no receptor-mediated chemotactic activity for HD-5 and HD-6 have been reported. Thus, receptors involved in chemoattraction triggered by defensins remain incompletely understood.

Attraction of immune cells to the site of injury can be indirectly mediated by HDPs by promoting expression and release of chemokines. HBD1–3 caused dose-dependent increases in secretion of chemokines IL-8, MCP-1, and mono-cyte inhibitory protein (MIP)-1-β by human peripheral blood mononuclear cells [94]. In bronchial epithelial cells, HNP1–3 caused dose-dependent increases in IL-8 synthesis via the P2Y6 receptor [80]. Likewise, HD5 induced expression of IL-8 and CCL20 (lymphocyte and neutrophil chemo attractant) in human intestinal (Caco2) cells [95] (Fig. 1).

After chemo-attracting immune cells to the site of insult, HDPs can regulate differentiation/activation of immune cells toward a particular cytokine profile, depending on cell type and inflammatory stimuli. For example, cathelicidin LL-37 promoted polarization of peripheral blood-derived monocytes toward an M1 phenotype in the presence of macrophage colony-stimulating factor (M-CSF), thereby secreting more IL-12p40 and less IL-10 upon exposure to LPS [96]. Similarly, LL-37 was transported to the nucleus in monocyte-derived dendritic cells (MDDCs) to promote their maturation by inducing expression of antigen-presenting human leukocyte antigen (HLA)-DR and costimulatory molecule cluster of differentiation (CD86) [97]. In addition, during psoriasis, LL-37 promoted its own uptake by plasmacytoid dendritic cells (pDCs) via interactions with self-DNA [98]. This DC induced differentiation and activation by LL-37, either alone or in combination with self-DNA, promoted expression of proinflammatory cytokines, including IL-12 and IFN-α [98,99].

Influx of neutrophils in the gut is essential to respond to infectious agents, such as *S. typhimurium* or *E. coli*. Cathelicidins seem to promote neutrophil-killing mechanisms. Incubation of neutrophils with LL-37 prolonged cell survival through FPRL-1 and P2X7 dependent caspase-3 activation [100]. Moreover, pretreatment of human and murine neutrophils with LL-37 increased production of reactive oxygen species (ROS) and promoted enhanced phagocytosis activity upon challenge with heat-inactivated *P. aeruginosa* or *S. aureus* [101]. Degranulation of mast cells is another key event in gut immunity, as mast cells release a mixture of compounds from their cytoplasmic granules, including histamine, proteoglycans, serotonin, and serine proteases, that trigger inflammatory defenses (e.g., increase permeability of the capillaries to allow leukocytes to engulf pathogens in an infected gut). Cathelicidins LL-37 degranulated mast cells and promoted secretion of cytokines IL-2, IL-4, IL-6, and IL-31 in a GPCR and PI3K dependent manner [102]. Cathelicidins also contributed to expression and release of cytokines by nonimmune cells.

Human bronchial epithelial (16HBE14) cells upon stimulation with LL-37 secreted IL-6 in an NF-κB activation-dependent manner [103]. Likewise, LL-37, in synergism with IL-17 and IL-22, dose dependently promoted IL-6 secretion by primary human keratinocytes [90].

Defensins are also effectors inducing differentiation, activation, and cytokine release by hematopoietic and nonhematopoietic cells. HBD2 (and to a lesser extent HBD1) promoted secretion of IL-6 by human peripheral blood mononuclear cells [94]. HBD3 induced expression of costimulatory molecules CD80, CD86, and CD40 on peripheral blood-derived human mDCs and monocytes, in a TLR-dependent fashion [77]. Likewise, HNPs simultaneously induced secretion of TNF-α and IL-1β protein and inhibited expression of IL-10 mRNA in human peripheral blood-derived monocytes [104]. HD5 also activated the NF-κB pathway to induce protein synthesis of IL-2 and TNF-α in intestinal epithelial (Caco2) cells and IFN-γ in peripheral blood-purified CD4+ T cells [95]. Taken together, HDPs functioned as linking molecules in gut defenses, promoting recruitment of effector cells and activating release of proinflammatory cytokines, essential host innate defenses in combating infectious enteric diseases.

3.2.2 Antiinflammatory Effects of HDPs

In addition to proinflammatory roles, HDPs have been associated with mechanisms to resolve damage, including promoting wound healing and epithelial restitution through chemoattraction of endothelial and epithelial cells. These processes are key for mitigating damage associated with persistent inflammation and reestablishing gut homeostasis. LL-37 promoted migration of colonic epithelial (Caco2) cells in a P2X7, EGFR, and p38MAPK activation-dependent manner [105]. In addition, LL-37 acted via ligand-mediated EGFR phosphorylation to induce wound healing and restitution in primary human keratinocytes [106]. Interestingly, by inducing release of IL-8, cathelicidins may also contribute to epithelial restitution. In addition to attracting neutrophils, IL-8 stimulated cell proliferation and migration of intestinal epithelial cells (HT-29 and Caco2) via binding CXCR1 (IL-8) receptor [107]. Moreover, IL-8 sent prosurvival signals to both fetal intestinal (H4) cells and mature (Caco2) cells via CXCR-2 (IL-8) receptors [108].

Suppression of immune cell activation and cytokine expression are among other regulatory mechanisms described for cathelicidins. Human cathelicidin LL-37 downregulated IFN-γ-meditated proinflammatory cytokine responses (TNF-α, Il-6 and IL-12) in human peripheral blood mononuclear cells (PBMCs), expression of maturation markers (CD80, CD86 and MHCII) on monocytes and dendritic cells, and proliferation and class switching in mouse splenic B-cells [109]. This suppressor role of cathelicidins has been also noted in response to bacterial incitants. Human or murine neutrophils exposed to LL-37 had reduced secretion of proinflammatory cytokines

IL-1β, IL-6, IL-8, and TNF-α in response to heat-inactivated *P. aeruginosa* or *S. aureus* [101]. More specifically, bacterial virulence factors seemed to modulate such cathelicidin suppression. Human or mouse macrophage-like cells in the presence of LL-37 had reduced expression of TNF-α and nitric oxide (NO) when stimulated with bacterial lipooligosaccharide or LPS [88,110]. Likewise, LL-37 caused dose-dependent increases in production of antiinflammatory IL-10 and counteracted LPS-induced expression of proinflammatory IL-6, TNF-α, IL-8, and MCP-1 cytokines in myeloid-derived dendritic cell (mDCs), pDCs, CD14+ monocytes, B-cell, and T-cells [111]. In other aspects of gut restitution, LL-37 induced expression of gut mucins (MUC1 and 2) in colonic epithelial (HT-29) cells via MEK1/2 activation [112]. Reciprocally, *Muc2*-deficient mice infected with *E. histolytica* had significantly lower expression of mouse cathelicidin (mCramp) compared to wild-type C57BL/6 mice [113].

Defensins also have antiinflammatory properties, including chemoattraction-dependent wound healing (proliferation and restitution) and induction of antiinflammatory cytokine production by immune and nonimmune cells. HBD2–4 promoted migration and IL-10 cytokine expression in normal human epidermal keratinocytes in a GPCR-dependent manner [114]. HBD3 (and mouse orthologue Defb14) downregulated LPS-induced TNF-α and IL-6 expression by mouse bone marrow-derived macrophages and human monocyte-derived macrophage, respectively [115]. In agreement, HBD3 inhibited LPS-mediated NF-κB activation and thus reduced expression of proinflammatory cytokines (e.g., IL-12p40) in bone marrow-derived mouse macrophages and mDCs [116], whereas activation of defensins seemed to regulate such immunosuppression. Mice with impaired intestinal α-defensins activation (MMP7$^{-/-}$) had higher baseline levels of proinflammatory IL-1β secretions in colon and cecum, compared to control, wild-type mice [117]. Moreover, to confirm that α-defensins were instrumental in controlling proinflammatory IL-1β secretion, mouse peritoneal macrophages isolated from C57BL/6 mice were stimulated with IL-1β inducers (LPS and ATP) and had reduced levels of pro- and cleaved IL-1β secretion in the presence of synthetic cryptdins 3 or 4 (murine homologues of human intestinal α-defensins) [117]. The effect of defensins may also rely on antimicrobial effects on the gut microbiome. HD5 overexpressing transgenic mice have a shift in the gut microbiome, and hence fewer proinflammatory and allergenic IL-17-producing Th17 cells in intestinal lamina propria compared to control wild-type mice [118]. In terms of mucus barrier restitution, HNP1–3 promoted mucin (MUC5B and MUC5AC) expression and proliferation in human pulmonary NCI-H292 epithelial cells [119]. Whether HDPs' regulate mucin expression in the gut remain unknown; however, in a recent study, *Muc2*-deficient mice (homo or heterozygous) had negligible basal expression of mBD14 (mouse orthologue of hBD3) compared to wild-type C57BL/6 mice [120].

3.3   Interactions Between HDPs and Pattern Recognition Receptors

PRRs are germline encoded receptors that recognize constitutive and conserved pathogen-associated molecular patterns (PAMPs) from microbes or by-derived products. Among PRRs, TLRs are well known by their ability to distinguish ligands from microbial surfaces or cytoplasm. Activation of TLRs eventually leads to production of inflammatory cytokines and costimulatory molecules, thus initiating innate and adaptive immune responses [121]. Whereas at least 10 functional TLR genes (TLR1–10) have been described in humans [122,123], mRNA genes from TLR1 to TLR9 have been identified in both small intestine and colon [124,125] (Table 2). In the small intestine, TLRs had maximal expression in the villi, with the crypt zone having only weak expression. In the large intestine, TLR expression was similarily distributed across crypts and the upper part of the epithelium [134]. Specific for TLR types, baseline mRNA and protein TLR2 and TLR4 were most abundant in intestinal epithelial cells in normal colons [125,135]. Moreover, TLR3 mRNA and protein were abundant in normal small intestine and colon, whereas TLR5 protein was expressed predominantly in the colon [126,136] (Table 2). At a cellular level, cytoplasmic protein expression of TLRs has been described in intestinal epithelial cells, with TLR8 and TLR9 occasionally present on cell membranes [134].

To differentially respond to signals from the lumen or interstitium, TLRs were determined to be asymmetrically dispersed in intrinsically polarized intestinal epithelial cells, with their apical and basolateral membrane compartments. In this regard, expression of TLR2 and TLR4 was restricted to the apical membrane on differentiated enterocytes [127]. In contrast, TLR5 was mostly expressed at the basolateral membrane of intestinal epithelial cells and TLR9 at both apical and basolateral membranes (Table 2). This distribution of TLRs makes gut epithelial cells broad sentinels, capable of detecting both luminal and systemic pathogens.

To maintain gut homeostasis, intestinal epithelial cells express low constitutive levels of TLR2 and TLR4 and are poorly responsive to bacterial ligands [125,137]. This hyporesponsiveness to TLR ligands in the gut epithelium is attributed to a physiological mechanism to prevent exaggerated responses to microbes that normally reside at intestinal surfaces. Downregulation of TLR surface expression was accompanied by upregulation of a TLR inhibitory molecule, the protein Tollip, and decreased phosphorylation of interleukin-1 receptor-associated kinase (IRAK) [125].

Under inflammatory process such as IBDs or infectious colitis, TLRs can detect the presence of infiltrating microbes or associated virulent components and initiate synthesis of proinflammatory cytokines. In fact, activation of tissue macrophages by TLRs is a main mechanism for increased production of macrophage-derived cytokines, that is, TNF-α, IL-1, and IL-6, salient characteristics of IBD [138,139]. Similarly, intestinal epithelial cells in IBDs express

TABLE 2 Host Defense Peptides Intertalk With Different Toll-Like Receptors (TLRs) in the Intestinal Epithelium

Toll-Like Receptor (TLR)	TLR Expression in Normal and Inflamed Intestinal Epithelial Cells	Induction of Host Defense Peptides by TLR Activation in Intestinal Epithelial Cells	Modulation of TLRs by Host Defense Peptides in Intestinal Epithelial Cells	Refs.
TLR1	RNA and protein expression	ND	ND	[124,125]
TLR2	RNA and protein expression at low levels and restricted to apical compartments. Gene expression is minimally altered in IBDs	Activation induces HBD2 in NFκB and AP-1-dependent pathway	ND	[124–130]
TLR3	Abundant RNA and protein expression in small intestine and colon. Gene expression is minimally altered in IBDs	Agonists reduce bacterial load and inflammation by promoting murine cathelicidin mCRAMP	ND	[124–126,129,131]
TLR4	RNA and protein expression at low levels and restricted to apical compartments. Transcriptional expression is increased in IBDs	Activation induces HBD2 in NFκB and AP-1 dependent pathway. Low HBD2 expression is accompanied by low TLR4/MD2 expression in necrotizing enterocolitis. *Tlr4* ko mice have reduced α-defensin expression	Human cathelicidins (LL-37) induce TLR4 RNA and protein synthesis on apical intestinal epithelium by MAPK activation, which results in increased CXCL8 secretion and epithelial antimicrobial defenses against *E. coli*	[124–128,130,132,132a]

TLR			
TLR5	RNA expression in small and large intestine and protein predominantly in colon. It is restricted to basolateral compartments. Gene expression is minimally altered in IBDs	ND	ND	[124–126,129]
TLR6	RNA and protein expression	ND	ND	[124,125]
TLR7	RNA and protein expression	ND	HBD2 downregulates TLR7 expression	[124,125,133]
TLR8	RNA and protein expression	ND	ND	[124,125]
TLR9	RNA and protein expression at apical and basolateral membranes. Gene expression is minimally altered in IBDs	*Tlr9* ko mice have reduced α-defensin expression. *Tlr9* ko mice show worsened colitis and reduced colonic mCRAMP expression treatment after DSS	Human cathelicidins (LL-37) block TLR9 RNA and protein induced by agonist on apical intestinal epithelium	[43,124–126,128,130,132a]

These ligand/receptor interations modulate either cathelicidins and defensins or TLRs and eventually, the inflammatory responses in the gut. ND, not determined.

higher levels of TLR4 than normal cells and may be an additional source of proinflammatory cytokines [126,140]. In contrast, gene expression of other TLRs, including TLR2, TLR3, TLR5, and TLR9, were minimally altered in the intestinal epithelium of patients with IBDs [126,129] (Table 2).

Expression of HDPs can be directly modulated by TLRs or indirectly, via microbes that activate PRRs [141]. This role of TLR activation for modulating HDP has been reported in several tissues/organs. For instance in the eye, stimulation of TLRs expressed on the ocular surface cells by specific agonists for TLR1, 2, 3, 4, 5, or 6 induced production of LL-37 and HBD2 [141]. This synthesis of HDPs protected the ocular surface against bacterial infections [141]. As a negative feedback mechanism, HDPs subsequently downregulated mRNA expression of some TLRs (TLR5, 7, and 9) in human corneal and conjunctival epithelial cells [141]. Similarly, in lung epithelial cells, stimulation of TLR2 with peptidoglycan (PGN) or lipopeptides upregulated HBD2 expression [142,143]. Coinfections with virus and bacteria synergistically enhanced HBD2 in airway epithelia, in part due to interactions of flagellin protein with TLR5, as TLR5 knockdown attenuated this synergistic increase of HBD2 [144]. Effects of TLR on HDP synthesis were correspondingly described in monocyte-derived macrophages during *Mycobacterium tuberculosis* infection, in which LL-37 expression was upregulated through activation of TLR2, TLR4, and TLR9 [145]. The specific importance of TLR9 in the gut cathelicidin expression has been well characterized. Bacterial *E. coli* DNA, a natural TLR9 ligand, increased gene LL-37 transcription in human primary monocytes via an ERK1/2-dependent signaling pathway [43]. Pretreatment with TLR9 inhibitor, a RNA transcription inhibitor (actinomycin D), or protein translation inhibitor (cycloheximide) significantly inhibited LL-37 expression induced by *E. coli* DNA, indicating that TLR9 activation caused de novo RNA transcription and protein synthesis of cathelicidins [43].

Regarding the biological significance of TLRs modulating HDPs in the gut, intrarectal administration of TLR3 agonists in mice innoculated with *S. flexneri* reduced intestinal bacterial load and mucosal inflammation by promoting expression of the murine cathelicidin mCramp in colonic epithelial cells [131]. Moreover, early combined β-defensin and TLR4 gut responses to *Campylobacter jejuni* lessened the acute inflammatory phase produced during subsequent exposures to the pathogen [146]. Further involvement of TLR9 in expression of cathelicidins was reported in *Tlr9* ko mice, which had worsened colitis and reduced colonic mCramp levels than wild-type after DSS treatment [43]. Microbial double-stranded RNA and its synthetic analog poly(I:C), ligands for TLR3, were tissue-specific inducers of cathelicidin mRNA and protein expression in murine and human intestinal epithelial cells, through phosphatidylinositol 3-kinase-protein kinase Cζ-Sp1 pathway [131]. This interaction between TLR3 agonist and cathelicidins alleviated clinical shigellosis in a mouse model [131]. Impairment of intestinal HDPs and TLRs interactions may be equally important in gut health. In necrotizing enterocolitis, a disease characterized by extremely low-birth-weight infants and immature immune defenses, squat HBD2 gene and protein expression were accompanied by low TLR4/MD2 expression [132]. Because HBD2 is at least partly regulated by TLR4 and MD2 signaling [147], inadequate responses to luminal bacteria due to TLR dysfunction as described in necrotizing enterocolitis, may prompt insufficient HDP activation, thereby predisposing these infants to enterocolitis [132] (Table 2).

There is a key role of NFκB among other regulatory proteins in the mechanisms by which TLR stimulation leads to HDP responses [148]. The expression of HBD2, upregulated by poly I:C stimulation through TLR3 in colonocytes, was specifically dependent on the NFκB signaling pathway [149]. Likewise, induction of HBD2 through activation of TLR4 (using agonist LPS) or TLR2 (using agonist PGN) in intestinal epithelial cells was abrogated by mutations in NFκB or related transcriptional factor AP-1 site within the HBD2 promoter [128]. In agreement, activation of both NF-κB and AP-1 was essential for complete *E. coli*-mediated induction of HBD2 mRNA in colonocytes [150]. Thus, transcription factors NF-κB and AP-1 were regarded as central regulators of epithelial HBD expression, likely by responding to TLR ligands [151]. More studies also indicated the importance of TLR co-factor MyD88 (TLR-associated downstream signaling adaptor) as a connector between TLRs and HDPs. Blocking MyD88 in intestinal epithelial cells increased susceptibility to infection by impairing synthesis of antimicrobial genes, including RegIIIγ and goblet cell-specific factors resistin-like molecule β (RELMβ) and mucin 2 (Muc2) and the reparative trefoil factor 3 (TFF3) [152]. Likewise, transgenic mice with impaired MyD88 function or protein loss had decreased production of α-defensins and RegIIIβ, as well as RELMβ and RegIIIγ, which increased susceptibility to DSS colitis [153–155]. Expression and secretion of ileal α-defensins by Paneth cells in response to bacterial stimuli [29] also depended on TLR-MyD88-signaling pathway [130]. Lastly, mice with genetic deficiencies in bacterial sensing ($Tlr2^{-/-}$, $Tlr4^{-/-}$, $Tlr9^{-/-}$, $Tlr2/4^{-/-}$, or $MyD88^{-/-}$) had significant reductions in expression of α-defensin genes, with $Tlr2^{-/-}/Tlr4^{-/-}$ double-deficient mice having the greates reductions in transcript levels for most α-defensin genes [130] (Table 2). Other interesting mechanisms between HDPs and TLR in intestinal epithelial cells were described in studies using protozoal stimuli, in which apical exosomes containing antimicrobial peptides (including cathelicidins and HBD2) were released through activation of TLR4 [156].

Although TLRs can modulate HDP expression, the latter may reciprocally act on inflammatory processes by modulating expression of TLRs (Table 2). In this regard, cathelicidins in synergy with Gram-negative *S. typhimurium* or *E. coli* or LPS increased transcription and protein synthesis of TLR4 when applied to the apical side of an intestinal epithelium [157]. These actions on TLR4 triggered by cathelicidins occurred mostly by activation of MAPK signaling pathways and resulted in increased CXCL8 chemokine secretion and epithelial antimicrobial defenses against *E. coli* [157]. Depending on cell types and cytokines in the surrounding milieu, HDPs may have also downregulatory effects on TLRs and synthesis of cytokines. In myeloid cells, LL-37 almost completely prevented release of TNF-α and other cytokines by human PBMC following stimulation with LPS and other TLR2/4 and TLR9 agonists [157a]. Likewise, coincubation of LL-37 with TLR ligands (LPS, lipoteichoic acid, and flagellin) suppressed TLR activation of monocyte-derived DCs, manifested as decreased release of IL-6, IL-12p70, and TNF-α and decreased surface expression of HLA-DR, CD80, CD83, CD86, and chemokine receptor CCR7 [157b]. Further studies also addressed the TLR suppressory effect of cathelicidins in DCs. Exposure of DCs to LL-37 during TLR4 LPS ligand exposure decreased proliferation of cocultured naive T cells and reduced production of IL-2 and IFN-γ [157b]. Mechanistically, the immunemodulatory role of LL-37 on TLRs in human monocytes targeted NFκB proinflammatory genes, including NFkappaB1 (p105/p50) and TNF-α-induced protein 2 (TNFAIP2) [157a]. Nuclear translocation of NF-κβ subunits p50 and p65 in LPS-treated monocytes was significantly reduced in the presence of LL-37 [157a]. Stimulation of colonocytes with HBD2 similarly downregulated mRNA expression of TLR7, IL-1R-associated kinase, neutrophil alphadefensins (1–3), and IL-8, but it upregulated C-AMP and NF κβ-p65 [133]. In addition, porcine β-defensin 2 (pBD2; orthologue of human β-defensin-1) given orally to weaned piglets challenged with enterotoxigenic *E. coli* decreased mRNA expression of TLR4 and related proinflammatory cytokines IL-1β, IL-8, and TNF-α in the jejunum epithelium and in the serum [158]. This immune-modulatory effect of pBD2 may have therapeutic benefits as an antimicrobial agent and gut growth enhancer in swine [158]. Thus, HDPs are key cell regulators in myeloid and nonmyeloid cells acting directly on the TLR-to-NF κβ pathway.

Regarding mechanisms by which HDPs modulate TLRs and inflammatory responses, cathelicidins and defensins are cationic peptides that can bind through electrostatic interaction to negatively charged molecules, including DNA [98], glycosaminoglycans [159], and mucin [160]. In this regard, LL-37 electrostatically interacted with the anionic lipid A portion of LPS and blocked binding of LPS to CD14(+) cells, thereby inhibiting expression of TNF-α [161]. However, under certain conditions, HDPs could facilitate uptake and signaling of TLR ligands. Cathelicidins promoted LPS-mediated activation of TLR4 in colonic epithelium and increased IL-8 synthesis [157]. In agreement, LL-37 facilitated uptake of LPS and its delivery to TLR4-containing intracellular lysosomes in airway epithelium [62]. This cathelicidin-dependent LPS internalization increased release of inflammatory cytokines IL-6 and IL-8 [62,162]. Mechanistically, LPS uptake regulated by cathelicidin was peptide-specific and involved endocytotic machinery, functional lipid rafts, and activation of MAPK/ERK and epidermal growth factor receptor signaling [62,162].

4   CONCLUSION

The pathogenesis of IBDs and infectious colitis comprises damaging proinflammatory gut responses and dysfunctional synthesis of HDPs, although whether it is a cause or an effect is unknown. Defensins and cathelicidins, through crosstalk with TLRs, are part of the pathogenesis of infectious and abiotic colitis. Although a full understanding is still in its infancy, HDPs seemed to regulate gut resistance to microbial colonization by triggering particular types of activation of immune cells, depending on the stressor and resulting cytokines. Clearly HDPs have potential to control both microbes and inflammation, together with a unique ability to differentially regulate immune responses depending on the tissue and inflammatory stressors. This dual role of HDPs is an excellent biological model to generate data that could be used to design evidence-based novel therapeutics. Most current therapeutics for IBDs are cytokine-based immunomodulators, whereas the primary focus of treatments for infectious diseases are conventional antibiotics, which are becoming increasingly unsustainable and ineffective due to emergence of antimicrobial resistance. Development of new therapeutics based on HDP biology is already underway. Synthetic cathelicidin derived from bovine neutrophil bactenecin (innate defense regulator peptide; IDR-1018) had beneficial therapeutic effects against IBDs [163]. Furthermore, it is expected that current and future research on peptides' structure, stability, efficacy, and targeted delivery with minimum side effects will advance clinical and practical uses of these peptides. Consequently, understanding HDP biological roles in gut defenses is opportunistic and important to aid in intervention strategies to cure or alleviate IBDs and infectious colitis.

ACKNOWLEDGMENTS

This work was supported by the Margaret Gunn Endowment for Animal Research (University of Calgary), NSERC Discovery Grant (RGPAS-2017-507827), and Alberta Agriculture and Forestry grant (2016E004R).

REFERENCES

[1] Navaneethan U, Giannella RA. Infectious colitis. Curr Opin Gastroenterol 2011;27:66–71.
[2] Papaconstantinou HT, Thomas JS. Bacterial colitis. Clin Colon Rectal Surg 2007;20:18–27.
[3] Ntuli ST, Malangu N, Alberts M. Causes of deaths in children under-five years old at a tertiary hospital in Limpopo province of South Africa. Global Journal of Health Science 2013;5:95–100.
[4] Kaplan GG. The global burden of IBD: from 2015 to 2025. Nat Rev Gastroenterol Hepatol 2015;12:720–7.
[5] Kaser A, Zeissig S, Blumberg RS. Inflammatory bowel disease. Annu Rev Immunol 2010;28:573–621.
[6] Fournier BM, Parkos CA. The role of neutrophils during intestinal inflammation. Mucosal Immunol 2012;5:354–66.
[7] Neurath MF. Cytokines in inflammatory bowel disease. Nat Rev Immunol 2014;14:329–42.

[8] Strober W, Fuss I. Proinflammatory cytokines in the pathogenesis of inflammatory bowel diseases. Gastroenterology 2011;140:1756–67.

[9] Fukata M, Arditi M. The role of pattern recognition receptors in intestinal inflammation. Mucosal Immunol 2013;6:451–63.

[10] Wirtz S, Neurath MF. Mouse models of inflammatory bowel disease. Adv Drug Deliv Rev 2007;59:1073–83.

[11] Galvez J. Role of Th17 cells in the pathogenesis of human IBD. ISRN Inflamm 2014;2014.

[12] Wu XW, Ji HZ, Wang FY. Meta-analysis of ciprofloxacin in treatment of Crohn's disease. Biomed Rep 2015;3:70–4.

[13] Hodges K, Gill R. Infectious diarrhea: cellular and molecular mechanisms. Gut Microbes 2010;1:4–21.

[14] Parronchi P, Romagnani P, Annunziato F, Sampognaro S, Becchio A, Giannarini L, Maggi E, Pupilli C, Tonelli F, Romagnani S. Type 1 T-helper cell predominance and interleukin-12 expression in the gut of patients with Crohn's disease. Am J Pathol 1997;150:823–32.

[15] Sarra M, Pallone F, Macdonald TT, Monteleone G. IL-23/IL-17 axis in IBD. Inflamm Bowel Dis 2010;16:1808–13.

[16] Fuss IJ, Heller F, Boirivant M, Leon F, Yoshida M, Fichtner-Feigl S, Yang Z, Exley M, Kitani A, Blumberg RS, Mannon P, Strober W. Nonclassical CD1d-restricted NK T cells that produce IL-13 characterize an atypical Th2 response in ulcerative colitis. J Clin Invest 2004;113:1490–7.

[17] Fuss IJ, Neurath M, Boirivant M, Klein JS, De La Motte C, Strong SA, Fiocchi C, Strober W. Disparate CD4+ lamina propria (LP) lymphokine secretion profiles in inflammatory bowel disease. Crohn's disease LP cells manifest increased secretion of IFN-gamma, whereas ulcerative colitis LP cells manifest increased secretion of IL-5. J Immunol 1996;157:1261–70.

[18] Khor B, Gardet A, Xavier RJ. Genetics and pathogenesis of inflammatory bowel disease. Nature 2011;474:307–17.

[19] Peyrin-Biroulet L, Deltenre P, De Suray N, Branche J, Sandborn WJ, Colombel JF. Efficacy and safety of tumor necrosis factor antagonists in Crohn's disease: meta-analysis of placebo-controlled trials. Clin Gastroenterol Hepatol 2008;6:644–53.

[20] Sorrentino D, Avellini C, Beltrami CA, Pasqual E, Zearo E. Selective effect of infliximab on the inflammatory component of a colonic stricture in Crohn's disease. Int J Colorectal Dis 2006;21:276–81.

[21] Spinelli A, Correale C, Szabo H, Montorsi M. Intestinal fibrosis in Crohn's disease: medical treatment or surgery? Curr Drug Targets 2010;11:242–8.

[21a] Zanetti M. The role of cathelicidins in the innate host defenses of mammals. Curr Issues Mol Biol 2005;7(2):179–96.

[22] Holani R, Shah C, Haji Q, Inglis GD, Uwiera RRE, Cobo ER. Proline-arginine rich (PR-39) cathelicidin: structure, expression and functional implication in intestinal health. Comp Immunol Microbiol Infect Dis 2016;49:95–101.

[23] Schauber J, Svanholm C, Termen S, Iffland K, Menzel T, Scheppach W, Melcher R, Agerberth B, Luhrs H, Gudmundsson GH. Expression of the cathelicidin LL-37 is modulated by short chain fatty acids in colonocytes: relevance of signalling pathways. Gut 2003;52:735–41.

[23a] Kosciuczuk EM, Lisowski P, Jarczak J, Strzalkowska N, Jozwik A, Horbanczuk J, Krzyzewski J, Zwierzchowski L, Bagnicka E. Cathelicidins: family of antimicrobial peptides. A review. Mol Biol Rep 2012;39(12):10957–70.

[24] Wang G. Human antimicrobial peptides and proteins. Pharmaceuticals (Basel) 2014;7:545–94.

[25] Ho S, Pothoulakis C, Koon HW. Antimicrobial peptides and colitis. Curr Pharm Des 2013;19:40–7.

[26] Tomasinsig L, Zanetti M. The cathelicidins structure, function and evolution. Curr Protein Pept Sci 2005;6:23–34.

[27] Cunliffe RN, Kamal M, Rose FR, James PD, Mahida YR. Expression of antimicrobial neutrophil defensins in epithelial cells of active inflammatory bowel disease mucosa. J Clin Pathol 2002;55:298–304.

[28] Wilson CL, Ouellette AJ, Satchell DP, Ayabe T, Lopez-Boado YS, Stratman JL, Hultgren SJ, Matrisian LM, Parks WC. Regulation of intestinal alpha-defensin activation by the metalloproteinase matrilysin in innate host defense. Science 1999;286:113–7.

[28a] Dann SM, Eckmann L. Innate immune defenses in the intestinal tract. Curr Opin Gastroenterol 2007;23(2):115–20.

[28b] Zhao A, Lu W, Leeuw E. Functional synergism of Human Defensin 5 and Human Defensin 6. Biochem Biophys Res Commun 2015;467(4):967–72.

[28c] Porter EM, Bevins CL, Ghosh D, Ganz T. The multifaceted Paneth cell. Cell Mol Life Sci 2002;59(1):156–70.

[29] Ayabe T, Satchell DP, Wilson CL, Parks WC, Selsted ME, Ouellette AJ. Secretion of microbicidal alpha-defensins by intestinal Paneth cells in response to bacteria. Nat Immunol 2000;1:113–8.

[30] Ganz T. Defensins: antimicrobial peptides of innate immunity. Nat Rev Immunol 2003;3:710–20.

[30a] Frye M, Bargon J, Lembcke B, Wagner TO, Gropp R. Differential expression of human alpha- and beta-defensins mRNA in gastrointestinal epithelia. Eur J Clin Investig 2000;30(8):695–701.

[31] Sperandio B, Regnault B, Guo J, Zhang Z, Stanley Jr SL, Sansonetti PJ, Pedron T. Virulent *Shigella flexneri* subverts the host innate immune response through manipulation of antimicrobial peptide gene expression. J Exp Med 2008;205:1121–32.

[32] Zaalouk TK, Bajaj-Elliott M, George JT, Mcdonald V. Differential regulation of beta-defensin gene expression during *Cryptosporidium parvum* infection. Infect Immun 2004;72:2772–9.

[33] Salzman NH, Chou MM, De Jong H, Liu L, Porter EM, Paterson Y. Enteric salmonella infection inhibits Paneth cell antimicrobial peptide expression. Infect Immun 2003;71:1109–15.

[34] Yi H, Hu W, Chen S, Lu Z, Wang Y. Cathelicidin-WA improves intestinal epithelial barrier function and enhances host defense against enterohemorrhagic *Escherichia coli* O157:H7 infection. J Immunol 2017;198:1696–705.

[35] Yoo JH, Ho S, Tran DH, Cheng M, Bakirtzi K, Kukota Y, Ichikawa R, Su B, Tran DH, Hing TC, Chang I, Shih DQ, Issacson RE, Gallo RL, Fiocchi C, Pothoulakis C, Koon HW. Anti-fibrogenic effects of the anti-microbial peptide cathelicidin in murine colitis-associated fibrosis. Cell Mol Gastroenterol Hepatol 2015;1:55–74. e1.

[36] Hing TC, Ho S, Shih DQ, Ichikawa R, Cheng M, Chen J, Chen X, Law I, Najarian R, Kelly CP, Gallo RL, Targan SR, Pothoulakis C, Koon HW. The antimicrobial peptide cathelicidin modulates *Clostridium difficile*-associated colitis and toxin A-mediated enteritis in mice. Gut 2013;62:1295–305.

[37] Salzman NH, Ghosh D, Huttner KM, Paterson Y, Bevins CL. Protection against enteric salmonellosis in transgenic mice expressing a human intestinal defensin. Nature 2003;422:522–6.

[38] Manko A, Motta JP, Cotton JA, Feener T, Oyeyemi A, Vallance BA, Wallace JL, Buret AG. Giardia co-infection promotes the secretion of antimicrobial peptides beta-defensin 2 and

trefoil factor 3 and attenuates attaching and effacing bacteria-induced intestinal disease. PLoS One 2017;12:.

[39] Tran DH, Wang J, Ha C, Ho W, Mattai SA, Oikonomopoulos A, Weiss G, Lacey P, Cheng M, Shieh C, Mussatto CC, Ho S, Hommes D, Koon HW. Circulating cathelicidin levels correlate with mucosal disease activity in ulcerative colitis, risk of intestinal stricture in Crohn's disease, and clinical prognosis in inflammatory bowel disease. BMC Gastroenterol 2017;17:63.

[40] Schauber J, Rieger D, Weiler F, Wehkamp J, Eck M, Fellermann K, Scheppach W, Gallo RL, Stange EF. Heterogeneous expression of human cathelicidin hCAP18/LL-37 in inflammatory bowel diseases. Eur J Gastroenterol Hepatol 2006;18:615–21.

[41] Hase K, Eckmann L, Leopard JD, Varki N, Kagnoff MF. Cell differentiation is a key determinant of cathelicidin LL-37/human cationic antimicrobial protein 18 expression by human colon epithelium. Infect Immun 2002;70:953–63.

[42] Wong CC, Zhang L, Li ZJ, Wu WK, Ren SX, Chen YC, Ng TB, Cho CH. Protective effects of cathelicidin-encoding *Lactococcus lactis* in murine ulcerative colitis. J Gastroenterol Hepatol 2012;27:1205–12.

[43] Koon HW, Shih DQ, Chen J, Bakirtzi K, Hing TC, Law I, Ho S, Ichikawa R, Zhao D, Xu H, Gallo R, Dempsey P, Cheng G, Targan SR, Pothoulakis C. Cathelicidin signaling via the Toll-like receptor protects against colitis in mice. Gastroenterology 2011;141:1852–63. e1–3.

[44] Raftery T, Martineau AR, Greiller CL, Ghosh S, Mcnamara D, Bennett K, Meddings J, O'sullivan M. Effects of vitamin D supplementation on intestinal permeability, cathelicidin and disease markers in Crohn's disease: results from a randomised double-blind placebo-controlled study. United Eur Gastroenterol J 2015;3:294–302.

[45] Wehkamp J, Wang G, Kubler I, Nuding S, Gregorieff A, Schnabel A, Kays RJ, Fellermann K, Burk O, Schwab M, Clevers H, Bevins CL, Stange EF. The Paneth cell alpha-defensin deficiency of ileal Crohn's disease is linked to Wnt/Tcf-4. J Immunol 2007;179:3109–18.

[46] Simms LA, Doecke JD, Walsh MD, Huang N, Fowler EV, Radford-Smith GL. Reduced alpha-defensin expression is associated with inflammation and not NOD2 mutation status in ileal Crohn's disease. Gut 2008;57:903–10.

[47] Tanabe H, Ayabe T, Maemoto A, Ishikawa C, Inaba Y, Sato R, Moriichi K, Okamoto K, Watari J, Kono T, Ashida T, Kohgo Y. Denatured human alpha-defensin attenuates the bactericidal activity and the stability against enzymatic digestion. Biochem Biophys Res Commun 2007;358:349–55.

[48] Cadwell K, Liu JY, Brown SL, Miyoshi H, Loh J, Lennerz JK, Kishi C, Kc W, Carrero JA, Hunt S, Stone CD, Brunt EM, Xavier RJ, Sleckman BP, Li E, Mizushima N, Stappenbeck TS, Virgin HWT. A key role for autophagy and the autophagy gene Atg16l1 in mouse and human intestinal Paneth cells. Nature 2008;456:259–63.

[49] Kaser A, Lee AH, Franke A, Glickman JN, Zeissig S, Tilg H, Nieuwenhuis EE, Higgins DE, Schreiber S, Glimcher LH, Blumberg RS. XBP1 links ER stress to intestinal inflammation and confers genetic risk for human inflammatory bowel disease. Cell 2008;134:743–56.

[50] Zilbauer M, Jenke A, Wenzel G, Goedde D, Postberg J, Phillips AD, Lucas M, Noble-Jamieson G, Torrente F, Salvestrini C, Heuschkel R, Wirth S. Intestinal alpha-defensin expression in pediatric inflammatory bowel disease. Inflamm Bowel Dis 2011;17:2076–86.

[51] Ramasundara M, Leach ST, Lemberg DA, Day AS. Defensins and inflammation: the role of defensins in inflammatory bowel disease. J Gastroenterol Hepatol 2009;24:202–8.

[52] Zilbauer M, Jenke A, Wenzel G, Postberg J, Heusch A, Phillips AD, Noble-Jamieson G, Torrente F, Salvestrini C, Heuschkel R, Wirth S. Expression of human beta-defensins in children with chronic inflammatory bowel disease. PLoS One 2010;5.

[53] Fahlgren A, Hammarstrom S, Danielsson A, Hammarstrom ML. Beta-defensin-3 and -4 in intestinal epithelial cells display increased mRNA expression in ulcerative colitis. Clin Exp Immunol 2004;137:379–85.

[54] Ishikawa C, Tanabe H, Maemoto A, Ito T, Watari J, Kono T, Fujiya M, Ashida T, Ayabe T, Kohgo Y. Precursor processing of human defensin-5 is essential to the multiple functions *in vitro* and *in vivo*. J Innate Immun 2010;2:66–76.

[55] KjÆR TMR, Andersen B, Brinch KS. Oral treatment of inflammatory bowel disease; 2014 [Google Patents].

[56] Kruse T, Mygind PH, Brinch KS, KjÆrulff S, Andersen B, KjÆR TMR. Treatment of inflammatory bowel diseases with mammal beta defensins; 2011 [Google Patents].

[57] Mailänder-Sánchez D, Kjaerulf S, Sidelmann Brinch K, Andersen B, Stange EF, Malek N, Nordkild P, Wehkamp J. DOP083 recombinant subcutaneous human beta-Defensin 2 (hBD2) ameliorates experimental colitis in different in vivo models. J Crohn's Colitis 2017;11(Suppl.1):S75–6.

[58] Nguyen LT, Haney EF, Vogel HJ. The expanding scope of antimicrobial peptide structures and their modes of action. Trends Biotechnol 2011;29:464–72.

[59] Murakami M, Dorschner RA, Stern LJ, Lin KH, Gallo RL. Expression and secretion of cathelicidin antimicrobial peptides in murine mammary glands and human milk. Pediatr Res 2005;57:10–5.

[60] Xia X, Zhang L, Wang Y. The antimicrobial peptide cathelicidin-BF could be a potential therapeutic for *Salmonella typhimurium* infection. Microbiol Res 2015;171:45–51.

[61] Sandgren S, Wittrup A, Cheng F, Jonsson M, Eklund E, Busch S, Belting M. The human antimicrobial peptide LL-37 transfers extracellular DNA plasmid to the nuclear compartment of mammalian cells via lipid rafts and proteoglycan-dependent endocytosis. J Biol Chem 2004;279:17951–6.

[62] Shaykhiev R, Sierigk J, Herr C, Krasteva G, Kummer W, Bals R. The antimicrobial peptide cathelicidin enhances activation of lung epithelial cells by LPS. FASEB J 2010;24:4756–66.

[63] Negroni A, Cucchiara S, Stronati L. Apoptosis, necrosis, and necroptosis in the gut and intestinal homeostasis. Mediators Inflamm 2015;2015:250762.

[64] Chromek M, Arvidsson I, Karpman D. The antimicrobial peptide cathelicidin protects mice from *Escherichia coli* O157:H7-mediated disease. PLoS One 2012;7.

[65] Mathew B, Nagaraj R. Antimicrobial activity of human alpha-defensin 6 analogs: insights into the physico-chemical reasons behind weak bactericidal activity of HD6 *in vitro*. J Pept Sci 2015;21:811–8.

[66] Schroeder BO, Ehmann D, Precht JC, Castillo PA, Kuchler R, Berger J, Schaller M, Stange EF, Wehkamp J. Paneth cell alpha-defensin 6 (HD-6) is an antimicrobial peptide. Mucosal Immunol 2015;8:661–71.

[67] Peyrin-Biroulet L, Beisner J, Wang G, Nuding S, Oommen ST, Kelly D, Parmentier-Decrucq E, Dessein R, Merour E, Chavatte P, Grandjean T, Bressenot A, Desreumaux P, Colombel JF, Desvergne B, Stange EF, Wehkamp J, Chamaillard M. Peroxisome proliferator-activated receptor gamma activation is required for maintenance of innate antimicrobial immunity in the colon. Proc Natl Acad Sci U S A 2010;107:8772–7.

[68] Carryn S, Schaefer DA, Imboden M, Homan EJ, Bremel RD, Riggs MW. Phospholipases and cationic peptides inhibit *Cryptosporidium parvum* sporozoite infectivity by parasiticidal and non-parasiticidal mechanisms. J Parasitol 2012;98:199–204.

[68a] Johansson J, Gudmundsson GH, Rottenberg ME, Berndt KD, Agerberth B. Conformation-dependent antibacterial activity of the naturally occurring human peptide LL-37. J Biol Chem 1998;273:3718–24.

[69] Wang YH, Xia JL, Wang WM, Yang BW, Cui JF, Wang XD, Fan J. TNFalpha induced IL-8 production through p38 MAPK- NF-kB pathway in human hepatocellular carcinoma cells. Zhonghua Gan Zang Bing Za Zhi 2011;19:912–6.

[70] Huang LC, Jean D, Mcdermott AM. Effect of preservative-free artificial tears on the antimicrobial activity of human beta-defensin-2 and cathelicidin LL-37 in vitro. Eye Contact Lens 2005;31:34–8.

[71] Wong JH, Ye XJ, Ng TB. Cathelicidins: peptides with antimicrobial, immunomodulatory, anti-inflammatory, angiogenic, anticancer and procancer activities. Curr Protein Pept Sci 2013;14:504–14.

[72] Girnita A, Zheng H, Gronberg A, Girnita L, Stahle M. Identification of the cathelicidin peptide LL-37 as agonist for the type I insulin-like growth factor receptor. Oncogene 2012;31:352–65.

[73] De Y, Chen Q, Schmidt AP, Anderson GM, Wang JM, Wooters J, Oppenheim JJ, Chertov O. LL-37, the neutrophil granule- and epithelial cell-derived cathelicidin, utilizes formyl peptide receptor-like 1 (FPRL1) as a receptor to chemoattract human peripheral blood neutrophils, monocytes, and T cells. J Exp Med 2000;192:1069–74.

[74] Zhang Z, Cherryholmes G, Chang F, Rose DM, Schraufstatter I, Shively JE. Evidence that cathelicidin peptide LL-37 may act as a functional ligand for CXCR2 on human neutrophils. Eur J Immunol 2009;39:3181–94.

[75] Subramanian H, Gupta K, Guo Q, Price R, Ali H. Mas-related gene X2 (MrgX2) is a novel G protein-coupled receptor for the antimicrobial peptide LL-37 in human mast cells: resistance to receptor phosphorylation, desensitization, and internalization. J Biol Chem 2011;286:44739–49.

[76] Brandenburg LO, Jansen S, Wruck CJ, Lucius R, Pufe T. Antimicrobial peptide rCRAMP induced glial cell activation through P2Y receptor signalling pathways. Mol Immunol 2010;47:1905–13.

[77] Funderburg N, Lederman MM, Feng Z, Drage MG, Jadlowsky J, Harding CV, Weinberg A, Sieg SF. Human-defensin-3 activates professional antigen-presenting cells via Toll-like receptors 1 and 2. Proc Natl Acad Sci U S A 2007;104:18631–5.

[78] Yang D, Chertov O, Bykovskaia SN, Chen Q, Buffo MJ, Shogan J, Anderson M, Schroder JM, Wang JM, Howard OM, Oppenheim JJ. Beta-defensins: linking innate and adaptive immunity through dendritic and T cell CCR6. Science 1999;286:525–8.

[79] Rohrl J, Yang D, Oppenheim JJ, Hehlgans T. Human beta-defensin 2 and 3 and their mouse orthologs induce chemotaxis through interaction with CCR2. J Immunol 2010;184:6688–94.

[80] Khine AA, Del Sorbo L, Vaschetto R, Voglis S, Tullis E, Slutsky AS, Downey GP, Zhang H. Human neutrophil peptides induce interleukin-8 production through the P2Y6 signaling pathway. Blood 2006;107:2936–42.

[81] Tjabringa GS, Ninaber DK, Drijfhout JW, Rabe KF, Hiemstra PS. Human cathelicidin LL-37 is a chemoattractant for eosinophils and neutrophils that acts via formyl-peptide receptors. Int Arch Allergy Immunol 2006;140:103–12.

[82] Kurosaka K, Chen Q, Yarovinsky F, Oppenheim JJ, Yang D. Mouse cathelin-related antimicrobial peptide chemoattracts leukocytes using formyl peptide receptor-like 1/mouse formyl peptide receptor-like 2 as the receptor and acts as an immune adjuvant. J Immunol 2005;174:6257–65.

[83] Niyonsaba F, Iwabuchi K, Someya A, Hirata M, Matsuda H, Ogawa H, Nagaoka I. A cathelicidin family of human antibacterial peptide LL-37 induces mast cell chemotaxis. Immunology 2002;106:20–6.

[84] Montreekachon P, Chotjumlong P, Bolscher JG, Nazmi K, Reutrakul V, Krisanaprakornkit S. Involvement of P2X(7) purinergic receptor and MEK1/2 in interleukin-8 up-regulation by LL-37 in human gingival fibroblasts. J Periodontal Res 2011;46:327–37.

[85] Wu WK, Wong CC, Li ZJ, Zhang L, Ren SX, Cho CH. Cathelicidins in inflammation and tissue repair: potential therapeutic applications for gastrointestinal disorders. Acta Pharmacol Sin 2010;31:1118–22.

[86] Bowdish DM, Davidson DJ, Speert DP, Hancock RE. The human cationic peptide LL-37 induces activation of the extracellular signal-regulated kinase and p38 kinase pathways in primary human monocytes. J Immunol 2004;172:3758–65.

[87] Mookherjee N, Lippert DN, Hamill P, Falsafi R, Nijnik A, Kindrachuk J, Pistolic J, Gardy J, Miri P, Naseer M, Foster LJ, Hancock RE. Intracellular receptor for human host defense peptide LL-37 in monocytes. J Immunol 2009;183:2688–96.

[88] Scott MG, Davidson DJ, Gold MR, Bowdish D, Hancock RE. The human antimicrobial peptide LL-37 is a multifunctional modulator of innate immune responses. J Immunol 2002;169:3883–91.

[89] Yu J, Mookherjee N, Wee K, Bowdish DM, Pistolic J, Li Y, Rehaume L, Hancock RE. Host defense peptide Ll-37, in synergy with inflammatory mediator IL-1beta, augments immune responses by multiple pathways. J Immunol 2007;179:7684–91.

[90] Chen X, Takai T, Xie Y, Niyonsaba F, Okumura K, Ogawa H. Human antimicrobial peptide LL-37 modulates proinflammatory responses induced by cytokine milieus and double-stranded RNA in human keratinocytes. Biochem Biophys Res Commun 2013;433:532–7.

[91] Soruri A, Grigat J, Forssmann U, Riggert J, Zwirner J. Beta-defensins chemoattract macro-phages and mast cells but not lymphocytes and dendritic cells: CCR6 is not involved. Eur J Immunol 2007;37:2474–86.

[92] Yang D, Chen Q, Chertov O, Oppenheim JJ. Human neutrophil defensins selectively chemoattract naive T and immature dendritic cells. J Leukoc Biol 2000;68:9–14.

[93] Territo MC, Ganz T, Selsted ME, Lehrer R. Monocyte-chemotactic activity of defensins from human neutrophils. J Clin Invest 1989;84:2017–20.

[94] Boniotto M, Jordan WJ, Eskdale J, Tossi A, Antcheva N, Crovella S, Connell ND, Gallagher G. Human beta-defensin 2 induces a vigorous cytokine response in peripheral blood mononuclear cells. Antimicrob Agents Chemother 2006;50:1433–41.

[95] Lu W, De Leeuw E. Pro-inflammatory and pro-apoptotic properties of human Defensin 5. Biochem Biophys Res Commun 2013;436(3):557–62.

[96] Van Der Does AM, Beekhuizen H, Ravensbergen B, Vos T, Ottenhoff TH, Van Dissel JT, Drijfhout JW, Hiemstra PS, Nibbering PH. LL-37 directs macrophage differentiation toward macrophages with a proinflammatory signature. J Immunol 2010;185:1442–9.

[97] Bandholtz L, Ekman GJ, Vilhelmsson M, Buentke E, Agerberth B, Scheynius A, Gudmundsson GH. Antimicrobial peptide LL-37 internalized by immature human dendritic cells alters their phenotype. Scand J Immunol 2006;63:410–9.

[98] Lande R, Gregorio J, Facchinetti V, Chatterjee B, Wang YH, Homey B, Cao W, Wang YH, Su B, Nestle FO, Zal T, Mellman I, Schroder JM, Liu YJ, Gilliet M. Plasmacytoid dendritic cells sense self-DNA coupled with antimicrobial peptide. Nature 2007;449:564–9.

[99] Davidson DJ, Currie AJ, Reid GS, Bowdish DM, Macdonald KL, Ma RC, Hancock RE, Speert DP. The cationic antimicrobial peptide LL-37 modulates dendritic cell differentiation and dendritic cell-induced T cell polarization. J Immunol 2004;172:1146–56.

[100] Nagaoka I, Tamura H, Hirata M. An antimicrobial cathelicidin peptide, human CAP18/LL-37, suppresses neutrophil apoptosis via the activation of formyl-peptide receptor-like 1 and P2X7. J Immunol 2006;176:3044–52.

[101] Alalwani SM, Sierigk J, Herr C, Pinkenburg O, Gallo R, Vogelmeier C, Bals R. The antimicrobial peptide LL-37 modulates the inflammatory and host defense response of human neutrophils. Eur J Immunol 2010;40:1118–26.

[102] Niyonsaba F, Ushio H, Hara M, Yokoi H, Tominaga M, Takamori K, Kajiwara N, Saito H, Nagaoka I, Ogawa H, Okumura K. Antimicrobial peptides human beta-defensins and cathelicidin LL-37 induce the secretion of a pruritogenic cytokine IL-31 by human mast cells. J Immunol 2010;184:3526–34.

[103] Pistolic J, Cosseau C, Li Y, Yu JJ, Filewod NC, Gellatly S, Rehaume LM, Bowdish DM, Hancock RE. Host defence peptide LL-37 induces IL-6 expression in human bronchial epithelial cells by activation of the NF-kappaB signaling pathway. J Innate Immun 2009;1:254–67.

[104] Chaly YV, Paleolog EM, Kolesnikova TS, Tikhonov II, Petratchenko EV, Voitenok NN. Neutrophil alpha-defensin human neutrophil peptide modulates cytokine production in human monocytes and adhesion molecule expression in endothelial cells. Eur Cytokine Netw 2000;11:257–66.

[105] Otte JM, Zdebik AE, Brand S, Chromik AM, Strauss S, Schmitz F, Steinstraesser L, Schmidt WE. Effects of the cathelicidin LL-37 on intestinal epithelial barrier integrity. Regul Pept 2009;156:104–17.

[106] Tokumaru S, Sayama K, Shirakata Y, Komatsuzawa H, Ouhara K, Hanakawa Y, Yahata Y, Dai X, Tohyama M, Nagai H, Yang L, Higashiyama S, Yoshimura A, Sugai M, Hashimoto K. Induction of keratinocyte migration via transactivation of the epidermal growth factor receptor by the antimicrobial peptide LL-37. J Immunol 2005;175:4662–8.

[107] Sturm A, Baumgart DC, D'heureuse JH, Hotz A, Wiedenmann B, Dignass AU. CXCL8 modulates human intestinal epithelial cells through a CXCR1 dependent pathway. Cytokine 2005;29:42–8.

[108] Maheshwari A, Lacson A, Lu W, Fox SE, Barleycorn AA, Christensen RD, Calhoun DA. Interleukin-8/CXCL8 forms an autocrine loop in fetal intestinal mucosa. Pediatr Res 2004;56:240–9.

[109] Nijnik A, Pistolic J, Wyatt A, Tam S, Hancock RE. Human cathelicidin peptide LL-37 modulates the effects of IFN-gamma on APCs. J Immunol 2009;183:5788–98.

[110] Zughaier SM, Shafer WM, Stephens DS. Antimicrobial peptides and endotoxin inhibit cytokine and nitric oxide release but amplify respiratory burst response in human and murine macrophages. Cell Microbiol 2005;7:1251–62.

[111] Mookherjee N, Hamill P, Gardy J, Blimkie D, Falsafi R, Chikatamarla A, Arenillas DJ, Doria S, Kollmann TR, Hancock RE. Systems biology evaluation of immune responses induced by human host defence peptide LL-37 in mononuclear cells. Mol Biosyst 2009;5:483–96.

[112] Tai EK, Wong HP, Lam EK, Wu WK, Yu L, Koo MW, Cho CH. Cathelicidin stimulates colonic mucus synthesis by up-regulating MUC1 and MUC2 expression through a mitogen-activated protein kinase pathway. J Cell Biochem 2008;104:251–8.

[113] Cobo ER, Kissoon-Singh V, Moreau F, Holani R, Chadee K. MUC2 mucin and butyrate contribute to the synthesis of the antimicrobial peptide cathelicidin in response to *Entamoeba histolytica* and DSS-induced colitis; 2017 [Infect Immun].

[114] Niyonsaba F, Ushio H, Nakano N, Ng W, Sayama K, Hashimoto K, Nagaoka I, Okumura K, Ogawa H. Antimicrobial peptides human beta-defensins stimulate epidermal keratinocyte migration, proliferation and production of proinflammatory cytokines and chemokines. J Invest Dermatol 2007;127:594–604.

[115] Semple F, Webb S, Li HN, Patel HB, Perretti M, Jackson IJ, Gray M, Davidson DJ, Dorin JR. Human beta-defensin 3 has immunosuppressive activity in vitro and in vivo. Eur J Immunol 2010;40:1073–8.

[116] Semple F, Macpherson H, Webb S, Cox SL, Mallin LJ, Tyrrell C, Grimes GR, Semple CA, Nix MA, Millhauser GL, Dorin JR. Human beta-defensin 3 affects the activity of pro-inflammatory pathways associated with MyD88 and TRIF. Eur J Immunol 2011;41:3291–300.

[117] Shi J, Aono S, Lu W, Ouellette AJ, Hu X, Ji Y, Wang L, Lenz S, Van Ginkel FW, Liles M, Dykstra C, Morrison EE, Elson CO. A novel role for defensins in intestinal homeostasis: regulation of IL-1beta secretion. J Immunol 2007;179:1245–53.

[118] Salzman NH, Hung K, Haribhai D, Chu H, Karlsson-Sjoberg J, Amir E, Teggatz P, Barman M, Hayward M, Eastwood D, Stoel M, Zhou Y, Sodergren E, Weinstock GM, Bevins CL, Williams CB, Bos NA. Enteric defensins are essential regulators of intestinal microbial ecology. Nat Immunol 2010;11:76–83.

[119] Aarbiou J, Verhoosel RM, Van Wetering S, De Boer WI, Van Krieken JH, Litvinov SV, Rabe KF, Hiemstra PS. Neutrophil defensins enhance lung epithelial wound closure and mucin gene expression in vitro. Am J Respir Cell Mol Biol 2004;30:193–201.

[120] Cobo ER, Kissoon-Singh V, Moreau F, Chadee K. Colonic MUC2 mucin regulates the expression and antimicrobial activity of beta-defensin 2. Mucosal Immunol 2015;8:1360–72.

[121] Janeway Jr CA, Medzhitov R. Innate immune recognition. Annu Rev Immunol 2002;20:197–216.

[122] Kubarenko A, Frank M, Weber AN. Structure-function relationships of Toll-like receptor domains through homology modelling and molecular dynamics. Biochem Soc Trans 2007;35:1515–8.

[123] Sabah-Ozcan S, Baser A, Olcucu T, Baris IC, Elmas L, Tuncay L, Eskicorapci S, Turk NS, Caner V. Human TLR gene family members are differentially expressed in patients with urothelial carcinoma of the bladder. Urol Oncol 2017;35(12):674:e11–674:e17.

[124] Abreu MT. Toll-like receptor signalling in the intestinal epithelium: how bacterial recognition shapes intestinal function. Nat Rev Immunol 2010;10:131–44.

[125] Otte JM, Cario E, Podolsky DK. Mechanisms of cross hyporesponsiveness to Toll-like receptor bacterial ligands in intestinal epithelial cells. Gastroenterology 2004;126:1054–70.

[126] Cario E, Podolsky DK. Differential alteration in intestinal epithelial cell expression of toll-like receptor 3 (TLR3) and TLR4 in inflammatory bowel disease. Infect Immun 2000;68:7010–7.

[127] Cario E, Brown D, Mckee M, Lynch-Devaney K, Gerken G, Podolsky DK. Commensal-associated molecular patterns induce selective toll-like receptor-trafficking from apical membrane to cytoplasmic compartments in polarized intestinal epithelium. Am J Pathol 2002;160:165–73.

[128] Vora P, Youdim A, Thomas LS, Fukata M, Tesfay SY, Lukasek K, Michelsen KS, Wada A, Hirayama T, Arditi M, Abreu MT. Beta-defensin-2 expression is regulated by TLR signaling in intestinal epithelial cells. J Immunol 2004;173:5398–405.

[129] Pedersen G, Andresen L, Matthiessen MW, Rask-Madsen J, Brynskov J. Expression of Toll-like receptor 9 and response to bacterial CpG oligodeoxynucleotides in human intestinal epithelium. Clin Exp Immunol 2005;141:298–306.

[130] Menendez A, Willing BP, Montero M, Wlodarska M, So CC, Bhinder G, Vallance BA, Finlay BB. Bacterial stimulation of the TLR-MyD88 pathway modulates the homeostatic expression of ileal Paneth cell alpha-defensins. J Innate Immun 2013;5:39–49.

[131] Ta A, Thakur BK, Dutta P, Sinha R, Koley H, Das S. Double-stranded RNA induces cathelicidin expression in the intestinal epithelial cells through phosphatidylinositol 3-kinase-protein kinase Czeta-Sp1 pathway and ameliorates shigellosis in mice. Cell Signal 2017;35:140–53.

[132] Jenke AC, Zilbauer M, Postberg J, Wirth S. Human beta-defensin 2 expression in Elbw infants with severe necrotizing enterocolitis. Pediatr Res 2012;72:513–20.

[132a] Marin M, Holani R, Shah CB, Odeon A, Cobo ER. Cathelicidin modulates synthesis of Toll-like Receptors (TLRs) 4 and 9 in colonic epithelium. Mol Immunol 2017;91:249–58.

[133] Stroinigg N, Srivastava MD. Modulation of toll-like receptor 7 and LL-37 expression in colon and breast epithelial cells by human beta-defensin-2. Allergy Asthma Proc 2005;26:299–309.

[134] Kinnebrew MA, Ubeda C, Zenewicz LA, Smith N, Flavell RA, Pamer EG. Bacterial flagellin stimulates Toll-like receptor 5-dependent defense against vancomycin-resistant *Enterococcus* infection. J Infect Dis 2010;201:534–43.

[135] Abreu MT, Vora P, Faure E, Thomas LS, Arnold ET, Arditi M. Decreased expression of Toll-like receptor-4 and MD-2 correlates with intestinal epithelial cell protection against dysregulated proinflammatory gene expression in response to bacterial lipopolysaccharide. J Immunol 2001;167:1609–16.

[136] Furrie E, Macfarlane S, Thomson G, Macfarlane GT, Microbiology & Gut Biology Group, and Tayside Tissue & Tumour Bank. Toll-like receptors-2, − 3 and −4 expression patterns on human colon and their regulation by mucosal-associated bacteria. Immunology 2005;115:565–74.

[137] Melmed G, Thomas LS, Lee N, Tesfay SY, Lukasek K, Michelsen KS, Zhou Y, Hu B, Arditi M, Abreu MT. Human intestinal epithelial cells are broadly unresponsive to Toll-like receptor 2-dependent bacterial ligands: implications for host-microbial interactions in the gut. J Immunol 2003;170:1406–15.

[138] Kolios G, Petoumenos C, Nakos A. Mediators of inflammation: production and implication in inflammatory bowel disease. Hepatogastroenterology 1998;45:1601–9.

[139] Pallone F, Monteleone G. Regulatory cytokines in inflammatory bowel disease. Aliment Pharmacol Ther 1996;10(Suppl 2):75–9 [discussion 80].

[140] Vamadevan AS, Fukata M, Arnold ET, Thomas LS, Hsu D, Abreu MT. Regulation of Toll-like receptor 4-associated MD-2 in intestinal epithelial cells: a comprehensive analysis. Innate Immun 2010;16:93–103.

[141] Redfern RL, Reins RY, Mcdermott AM. Toll-like receptor activation modulates antimicrobial peptide expression by ocular surface cells. Exp Eye Res 2011;92:209–20.

[142] Hertz CJ, Wu Q, Porter EM, Zhang YJ, Weismuller KH, Godowski PJ, Ganz T, Randell SH, Modlin RL. Activation of Toll-like receptor 2 on human tracheobronchial epithelial cells induces the antimicrobial peptide human beta defensin-2. J Immunol 2003;171:6820–6.

[143] Homma T, Kato A, Hashimoto N, Batchelor J, Yoshikawa M, Imai S, Wakiguchi H, Saito H, Matsumoto K. Corticosteroid and cytokines synergistically enhance toll-like receptor 2 expression in respiratory epithelial cells. Am J Respir Cell Mol Biol 2004;31:463–9.

[144] Arnason JW, Murphy JC, Kooi C, Wiehler S, Traves SL, Shelfoon C, Maciejewski B, Dumonceaux CJ, Lewenza WS, Proud D, Leigh R. Human beta-defensin-2 production upon viral and bacterial co-infection is attenuated in COPD. PLoS One 2017;12.

[145] Rivas-Santiago B, Hernandez-Pando R, Carranza C, Juarez E, Contreras JL, Aguilar-Leon D, Torres M, Sada E. Expression of cathelicidin LL-37 during *Mycobacterium tuberculosis* infection in human alveolar macrophages, monocytes, neutrophils, and epithelial cells. Infect Immun 2008;76:935–41.

[146] Sung J, Morales W, Kim G, Pokkunuri V, Weitsman S, Rooks E, Marsh Z, Barlow GM, Chang C, Pimentel M. Effect of repeated *Campylobacter jejuni* infection on gut flora and mucosal defense in a rat model of post infectious functional and microbial bowel changes. Neurogastroenterol Motil 2013;25:529–37.

[147] Gariboldi S, Palazzo M, Zanobbio L, Selleri S, Sommariva M, Sfondrini L, Cavicchini S, Balsari A, Rumio C. Low molecular weight hyaluronic acid increases the self-defense of skin epithelium by induction of beta-defensin 2 via TLR2 and TLR4. J Immunol 2008;181:2103–10.

[148] Tsianos EV, Katsanos K. Do we really understand what the immunological disturbances in inflammatory bowel disease mean? World J Gastroenterol 2009;15:521–5.

[149] Omagari D, Takenouchi-Ohkubo N, Endo S, Ishigami T, Sawada A, Moro I, Asano M, Komiyama K. Nuclear factor kappa B plays a pivotal role in polyinosinic-polycytidylic acid-induced expression of human beta-defensin 2 in intestinal epithelial cells. Clin Exp Immunol 2011;165:85–93.

[150] Wehkamp J, Harder J, Wehkamp K, Wehkamp-Von Meissner B, Schlee M, Enders C, Sonnenborn U, Nuding S, Bengmark S, Fellermann K, Schroder JM, Stange EF. NF-kappaB- and AP-1-mediated induction of human beta defensin-2 in intestinal epithelial cells by Escherichia coli Nissle 1917: a novel effect of a probiotic bacterium. Infect Immun 2004;72:5750–8.

[151] Steubesand N, Kiehne K, Brunke G, Pahl R, Reiss K, Herzig KH, Schubert S, Schreiber S, Folsch UR, Rosenstiel P, Arlt A. The expression of the beta-defensins hBD-2 and hBD-3 is differentially regulated by NF-kappaB and MAPK/AP-1 pathways in an *in vitro* model of *Candida esophagitis*. BMC Immunol 2009;10:36.

[152] Bhinder G, Stahl M, Sham HP, Crowley SM, Morampudi V, Dalwadi U, Ma C, Jacobson K, Vallance BA. Intestinal epithelium-specific MyD88 signaling impacts host susceptibility to infectious colitis by promoting protective goblet cell and antimicrobial responses. Infect Immun 2014;82:3753–63.

[153] Brandl K, Plitas G, Schnabl B, Dematteo RP, Pamer EG. MyD88-mediated signals induce the bactericidal lectin RegIII gamma and protect mice against intestinal *Listeria monocytogenes* infection. J Exp Med 2007;204:1891–900.

[154] Gong J, Xu J, Zhu W, Gao X, Li N, Li J. Epithelial-specific blockade of MyD88-dependent pathway causes spontaneous small intestinal inflammation. Clin Immunol 2010;136:245–56.

[155] Vaishnava S, Behrendt CL, Ismail AS, Eckmann L, Hooper LV. Paneth cells directly sense gut commensals and maintain homeostasis at the intestinal host-microbial interface. Proc Natl Acad Sci U S A 2008;105:20858–63.

[156] Hu G, Gong AY, Roth AL, Huang BQ, Ward HD, Zhu G, Larusso NF, Hanson ND, Chen XM. Release of luminal exosomes contributes to TLR4-mediated epithelial antimicrobial defense. PLoS Pathog 2013;e1003261:9.

[157] Marin M, Holani R, Shah CB, Odeon A, Cobo ER. Cathelicidin modulates synthesis of Toll-like receptors (TLRs) 4 and 9 in colonic epithelium. Mol Immunol 2017;91:249–58.

[157a] Mookherjee N, Brown KL, Bowdish DM, Doria S, Falsafi R, Hokamp K, Roche FM, Mu R, Doho GH, Pistolic J, Powers JP, Bryan J, Brinkman FS, Hancock RE. Modulation of the TLR-mediated inflammatory response by the endogenous human host defense peptide LL-37. J Immunol 2006;176(4):2455–64.

[157b] Kandler K, Shaykhiev R, Kleemann P, Klescz F, Lohoff M, Vogelmeier C, Bals R. The anti-microbial peptide LL-37 inhibits the activation of dendritic cells by TLR ligands. Int Immunol 2006;18(12):1729–36.

[158] Tang Z XL, Shi B, Deng H, Lai X, Liu J, Sun Z. Oral administration of synthetic porcine beta-defensin-2 improves growth performance and cecal microbial flora and down-regulates the expression of intestinal toll-like receptor-4 and inflammatory cytokines in weaned piglets challenged with enterotoxigenic Escherichia coli. Anim Sci J 2006;87:1258–66.

[159] Baranska-Rybak W, Sonesson A, Nowicki R, Schmidtchen A. Glycosaminoglycans inhibit the antibacterial activity of LL-37 in biological fluids. J Antimicrob Chemother 2006;57:260–5.

[160] Felgentreff K, Beisswenger C, Griese M, Gulder T, Bringmann G, Bals R. The antimicrobial peptide cathelicidin interacts with airway mucus. Peptides 2006;27:3100–6.

[161] Nagaoka I, Hirota S, Niyonsaba F, Hirata M, Adachi Y, Tamura H, Heumann D. Cathelicidin family of antibacterial peptides CAP18 and CAP11 inhibit the expression of TNF-alpha by blocking the binding of LPS to CD14(+) cells. J Immunol 2001;167:3329–38.

[162] Tjabringa GS, Aarbiou J, Ninaber DK, Drijfhout JW, Sorensen OE, Borregaard N, Rabe KF, Hiemstra PS. The antimicrobial peptide LL-37 activates innate immunity at the airway epithelial surface by transactivation of the epidermal growth factor receptor. J Immunol 2003;171:6690–6.

[163] Hancock RE, Haney EF, Gill EE. The immunology of host defence peptides: beyond antimicrobial activity. Nat Rev Immunol 2016;16:321–34.

Chapter 8

Conclusion

Chi Hin Cho

Department of Pharmacology, School of Pharmacy, Southwest Medical University, Luzhou, China

The gastrointestinal (GI) tract is bound by a layer of epithelial cells covered by a physical barrier composing mainly dense mucus glycoproteins secreted from goblet cells. Furthermore, this epithelial defense is strengthened by tight junctions dictated by a paracellular pathway. This limits the host from the invasion of pathogens, toxins, intestinal microbiota, and dietary antigens attacking the mucosal proper, and so maintains mucosal integrity and health.

There are studies showing that aberrant expression of antimicrobial peptides (AMPs) in epithelial tissues is associated with barrier dysfunction in the GI tract. Moreover, any microbes, such as bacteria or any other pathogens from the lumen invading the mucosa are largely removed by the barrier and the innate immune system. All these self-defense domains are partly contributed to by well-defined AMPs found in the GI epithelium, not only through their unique antimicrobial action but also by their immunomodulatory activity and mucus secretion ability. These are the most effective host-defensive mechanisms operating in the GI mucosa. These physical barriers and immune responses abrogate bacterial translocation, inflammatory responses, and possibly the onset of various enteric and systemic disorders in human body.

Indeed, in recent decades, AMPs had been well studied in different organs and, in particular, in the GI tract. They exert multiple pleotropic effects on innate and adaptive immune responses. In addition, AMPs have a wide range of biological and pharmacological actions against bacterial infection and other significant pathogenic changes, such as inflammation and carcinogenesis. Interestingly, their antibacterial and anticancer actions are different from traditional drugs, with inherently unique mechanisms which have the advantage of not developing drug resistance after treatment. AMPs also inhere a unique mucosal-repairing property to reengineer ulcerous tissues back to a normal and functional mucosa.

The current documentation summarizes the various actions, either physiological or pharmacological, and the pivotal role played by AMPs, in particular, cathelicidin and defensins, in keeping the homeostasis in the microenvironment and health of the GI mucosa. It also covers a wide range of research areas from

Antimicrobial Peptides in Gastrointestinal Diseases. https://doi.org/10.1016/B978-0-12-814319-3.00008-8
Copyright © 2018 Chi Hin Cho. Published by Elsevier Ltd. All rights reserved.

the normal host defense behavior to pathogenic disorders, ranging from infection and inflammation and, further, to cancer in the gut.

In this book, we have multiple international members across the globe sharing their experiences and views in unveiling the relationship between AMP and different GI disorders. We have Dr. R. Bucki and his team from Poland who provide an overview on the balance between intestinal cells and microbes, and discuss the molecular mechanisms governing AMP production during homeostasis and GI disorders. They also propose various AMP inducers and their therapeutic values for the treatment of various GI diseases. Dr. L. Zhang and her group from Hong Kong review the host microbiota and focus on the functions of AMPs and the dynamic interaction of AMPs with commensal microbiota and pathogens. They maintain a homeostatic relationship between the host and its colonizing microbiota and play an essential part in modulating health and disease. From the USA, Dr. H.W. Koon and his associates have additional views on the biological actions of AMPs. They comment that the antimicrobial functions of AMPs are not necessarily the most important mechanism in the modulation of GI diseases. Instead, they may promote certain protective microbial species and modulate innate immune responses. AMPs may serve as biomarkers of different GI diseases because their expression is often altered during the pathogenic changes of those disorders. AMPs may even directly modulate the progression of obesity and metabolic diseases. Dr. J. Shen and Dr. Z.G. Xiao from China give a general overview on the pharmacological actions of cathelicidin on different diseases in animals. It protects against *Helicobacter pylori* infection and its associated gastritis, inhibits gastric cancer growth, and promotes gastric ulcer healing in mice and rats. It can also alleviate ulcerative colitis and perhaps also inhibit colon cancer growth in animals. Again, from the United States, Dr. L. Ma and her partners put forward their argument that human cathelicidin and defensins could have both pro- and anti-tumor effects on the GI tract. AMPs promote tumor growth at low concentrations but inhibit tumor proliferation in high doses. Various antitumor mechanisms are considered, which include membrane disruption, mitochondria damage, regulation of microRNA and gene transcription, alteration of cancer cell metabolism, and changes in the host immune system. AMPs also have unique anticancer properties. They overshadow other classical anticancer drugs, by escaping from the strategies used by cancer cells to develop drug-resistance mechanisms after chronic treatment. Dr. Lucinda Furci and Dr. Massimiliano Secchi from Italy give a detailed overview on the mechanisms of antibacterial action against different bacterial infections, in particular, the colonization of *Clostridium difficile* and its related disorders in the intestinal mucosa. They also comment on the structural features and locations of different AMPs in the GI tract and also from microbiota and immune cells. In this regard, the therapeutic potentials of defensins on *Salmonella typhimurium* and *C. difficile* infections are also discussed. Dr. E.R. Cobo and his group from Canada uncover the protective roles of cathelicidin and defensins, which promote epithelial defense, regulate inflammatory responses and mobilize leukocytes across the intestine for microbial elimination. This could unveil how the innate system can restore gut homeostasis and avoid inflammation associated with pathogen persistence in colons.

Indeed, research in AMPs is still in its rudimentary stage. There are more questions than answers in understanding the exact roles played by these peptides in different diseases. It is still not exactly known why they have such diversified biological and pharmacological actions against various kinds of disorders with completely different etiologies exacerbated in the GI tract. Nevertheless, they carry unique mechanisms of action different from the classical drugs currently used for these diseases. It is envisaged that AMPs will have great potentials to be developed as excellent therapeutic alternatives in the treatment of all these mucosal diseases, especially in the cases of multidrug resistance found in bacteria and cancer cells during the course of treatment against infection and carcinogenesis in human body.

Index

Note: Page numbers followed by *f* indicate figures and *t* indicate tables.